PREDICTING ROAD QUALITY AND ACCESSIBILITY INDICATORS THROUGH CONVENTIONAL AND INNOVATIVE METHODS

DECEMBER 2025

ASIAN DEVELOPMENT BANK

ADB

6 ADB Avenue, Mandaluyong City, 1550 Metro Manila, Philippines
Tel +63 2 8632 4444; Fax +63 2 8636 2444
www.adb.org

ISBN 978-92-9277-552-0 (print); 978-92-9277-553-7 (PDF); 978-92-9277-554-4 (ebook)
Publication Stock No. TCS250507-2
DOI: http://dx.doi.org/10.22617/TCS250507-2

Notes:
In this publication, "$" refers to United States dollars.
ADB recognizes "China" as the People's Republic of China.

Cover design by Mike Cortes.

Contents

Tables, Figures, and Boxes

Tables

Figures

Boxes

Foreword

Road quality monitoring plays a crucial role in ensuring safe, reliable, and efficient transportation systems. These, in turn, drive economic development, promote social inclusion, and improve access to essential services. By regularly assessing road conditions, governments and agencies can make informed, data-driven decisions to prioritize maintenance, allocate resources effectively, and prevent costly deterioration. Such monitoring also contributes to disaster resilience and ensures that rural and underserved communities remain connected. These are key components of equitable and sustainable development.

The need for reliable infrastructure to drive development is recognized in the Sustainable Development Goal (SDG) 9.1.1, which emphasizes universal access to all-season roads and is measured through the Rural Access Index (RAI). Collecting accurate RAI data remains a persistent challenge though, particularly in remote areas. This is due to financial limitations, accessibility issues, and human resource constraints, all factors that contribute to significant data gaps. Advances in machine learning and artificial intelligence offer promising, cost-effective alternatives for data collection, helping to bridge these gaps, enabling more inclusive and strategic infrastructure planning.

To further these efforts, the Japan Fund for Prosperous and Resilient Asia and the Pacific (JFPR), in partnership with the Asian Development Bank (ADB), has emerged as a key driver of sustainable rural development in the region. By addressing pressing infrastructure and connectivity needs, JFPR mobilizes resources and technical expertise to enhance the resilience and accessibility of rural areas. Its support for innovative, community-based solutions brings rural populations closer to essential services and opportunities, improving the overall quality of life for millions across Asia and the Pacific.

This JFPR-supported publication provides an overview of different approaches to monitoring road quality. It serves as a supplement to the *Guidebook on Machine Learning Techniques for Road Quality Monitoring* published in March 2025. The guidebook offers a comprehensive guide to implementing machine learning methods. This technical study focuses on the rationale and introduces core concepts related to monitoring road quality and its corresponding metrics. In particular, it includes a review of traditional methods for collecting road quality data, and a discussion on innovative technologies, such as using geospatial data and machine learning techniques. These can be further explored for more granular and timely data collection.

This publication was written by H. R. Pasindu, Arturo Martinez Jr., Joseph Albert Niño Bulan, Ron Lester Durante, Jayzon Mag-Atas, and Oshean Lee Garonita, under the overall direction of Elaine Tan and Jiang Yi. Data and valuable technical input were supplied by the Philippine Statistics Authority and the Department of Public Works and Highways. Thailand's National Statistical Office, Department of Highways, Department of Land Transport, and Department of Rural Roads also contributed. There was also input from participants in some events organized under this project, including a workshop with Yasuyuki Sawada from the University of Tokyo. Jiwoon Kang from ADB's transport sector and Poramintr Kingphakorn from ADB's Thailand Resident Mission provided feedback. Copyediting and manuscript editing were performed by Melanie Kelleher and Sean Crowley, respectively. Typesetting was conducted by Joe Mark Ganaban and the cover was designed by Mike Cortes. Proofreading was conducted by Monina Gamboa. Administrative support was provided by Rose Anne Dumayas and Iva Sebastian-Samaniego.

This publication aims to provide readers with a foundational understanding of both traditional and innovative approaches to road quality monitoring. It highlights how the adoption of emerging technologies and data-driven solutions can accelerate progress toward a more connected, resilient, and inclusive future for all.

Albert Park
Chief Economist and Director General
Economic Research and Development Impact Department
Asian Development Bank

Abbreviations

ADB	Asian Development Bank
ANN	artificial neural network
ASTM	American Standard of Testing and Materials
ATO	Asian Transport Observatory; Asian Transport Outlook (ATO applies to either)
CNN	convolutional neural network
DMC	developing member country
FAO	Food and Agriculture Organization
GPR	ground penetrating radar
IRI	international roughness index
km	kilometer
m/km	meters per kilometer
mm	millimeter
MLA	machine learning algorithm
OSM	OpenStreetMap
PSR	present serviceability rating
RAI	rural accessibility index
ReCAP	Research for Community Access Partnership
RTRRMS	response type road roughness measuring system
SDG	Sustainable Development Goal
TG	task group (with number after it)
UN	United Nations

I Background

Introduction

It is difficult to imagine a world without roads; no highways connecting cities, no rural paths linking villages to markets, schools, or hospitals. Roads are more than just strips of asphalt, they are lifelines of development. Decades of research have shown that transport infrastructure fuels productivity and economic output. Research clearly shows how public investment in infrastructure boosts national productivity (Aschauer 1989; Organisation for Economic Co-operation and Development 2002; World Bank 2019). Such investment also acts as capital inputs, empowering the private sector to thrive (Gramlich 1994). In developing economies, rural road investment has proved especially effective in delivering greater poverty reduction than other infrastructure types (Fan and Chan-Kang 2005). Improving road quality can accelerate economic growth and reduce income inequality too (Calderón and Servén 2008). The Asian Transport Observatory (ATO 2025) estimated that the investment need for transport infrastructure in the Asia and Pacific region will grow from $750 billion annually between 2000 and 2020 to $2.7 trillion annually between 2020 and 2035 with more than 50% allocated to the construction, maintenance, and climate proofing of road infrastructure.

Then came the coronavirus disease (COVID-19) pandemic in 2020. It reshaped how we move, literally. With physical distancing a global priority, commuting became a challenge. At that time, four possible post-pandemic transport scenarios, including a shift toward private and active transport modes like walking and cycling, had been outlined (ADB 2022). With the number of cars worldwide projected to hit 2 billion by 2040 (Smith, 2016), the urgency to rethink transport systems has never been greater.

Sustainable transport is no longer a luxury—it is a necessity. This means:

(i) promoting public transit;
(ii) creating safe, high-quality infrastructure for walking and cycling;
(iii) accelerating the adoption of electric vehicles; and
(iv) designing and delivering green infrastructure.

Figure 1: Estimated Transport Infrastructure Investment Needs (Construction, Maintenance, Climate-Proofing), by Sector
($ billion)

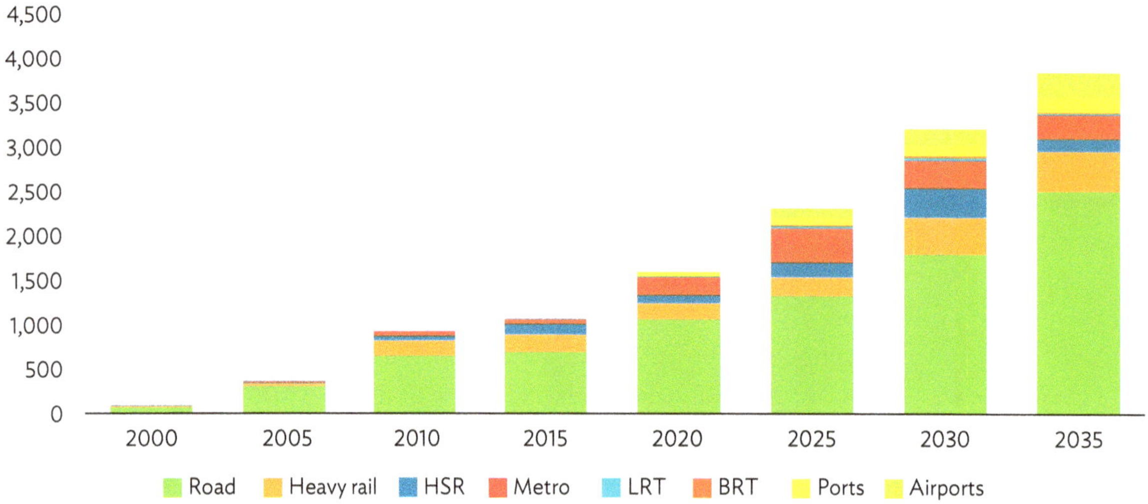

BRT = bud rapid transit, HSR = high-speed railway, LRT = light rail transit.
Source: Asian Transport Observatory. 2025. *Asia and the Pacific's Transport Infrastructure and Investment Outlook 2035.* https://asiantransportobservatory.org/analytical-outputs/asia-transport-infrastructure-investment-needs/.

Figure 2: Estimated Transport Infrastructure Investment Needs per Subregion, 2020 to 2035
(%)

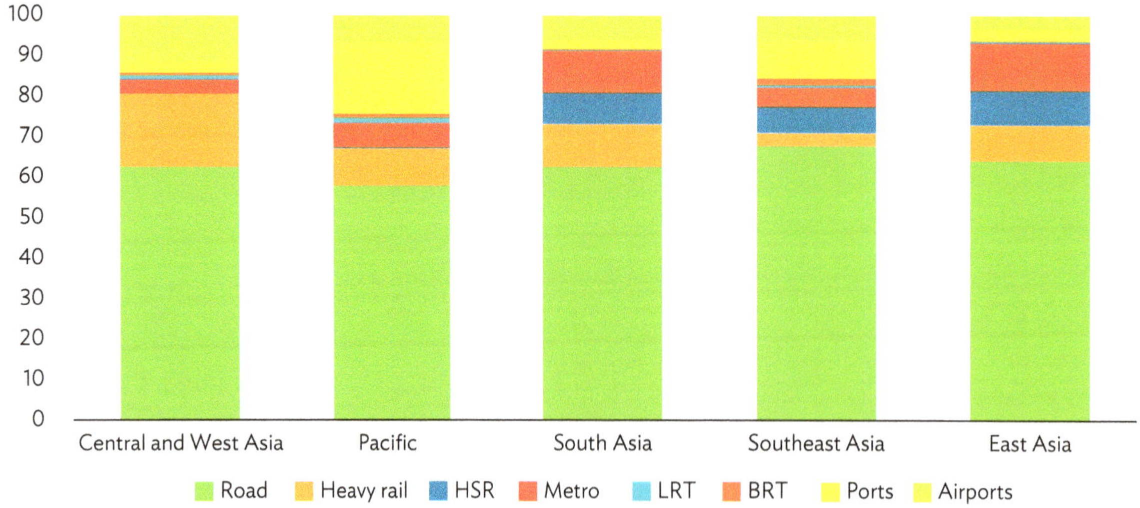

BRT = bud rapid transit, HSR = high-speed railway, LRT = light rail transit.
Source: Asian Transport Observatory. 2025. *Asia and the Pacific's Transport Infrastructure and Investment Outlook 2035.* https://asiantransportobservatory.org/analytical-outputs/asia-transport-infrastructure-investment-needs/.

Box 1: Road Quality as a Catalyst for Poverty Reduction and Economic Development—Insights from the Philippines and Thailand

In both the Philippines and Thailand, governments have long recognized that road quality is not just about infrastructure, it is about impact. It is about connecting people to jobs, education, healthcare, and markets. When roads deteriorate, so do those critical links.

In the Philippines, this vision is embedded in the Philippine Development Plan, which prioritizes both the construction and maintenance of road networks. With over 34,000 kilometers (km) of national roads, the Department of Public Works and Highways (DPWH) follows a maintenance cycle designed to preserve serviceability and avoid costly reconstruction. Yet environmental factors like rainfall and traffic volume continue to challenge road durability.

Thailand has prioritized extending its road network into remote and rugged terrain. But with expansion comes further complexity, particularly into more challenging regions. Monitoring road quality and evaluating its spillover effects has proven difficult. Budget and staffing constraints have led to delays in scheduled maintenance, and predictive models of deterioration still need refinement.

Consequences for poverty and development

In the Philippines, a study by Lawrence Dacuycuy and colleagues used spatial data, road quality information from the DPWH, satellite-based poverty maps from ADB, and household surveys to explore the link between roads and poverty. The findings were striking:

1. Poverty rates are converging locally, with poorer areas catching up with wealthier ones.
2. Road quality influences poverty outcomes, especially when analyzed at island level.
3. Spillover effects matter; districts surrounded by wealthier neighbors tend to perform better.
4. Human capital and local amenities may also play a role, though further data is needed.

In Thailand, Puttanapong and team proposed a contemporary method that uses satellite imagery and machine learning, neural networks, random forests, and support vector regression, to monitor road quality and assess localized impacts. Their models outperformed traditional regression methods and showed a clear link: paved roads boost economic growth, while poor road conditions (measured by international roughness index [IRI]) restrict growth.

Source: Authors.

But sustainability alone is not enough. Resilience must be built into the system. Transport infrastructure must be ready to withstand pandemics, floods, earthquakes, and other shocks. ADB (2022) has emphasized that advanced technologies like big data and artificial intelligence, paired with digital inclusion, can transform how we prepare for and respond to these challenges.

The evidence is clear: expanding road networks improves market access, boosts regional gross domestic product, and increases employment and population (Adler et al. 2020). These benefits are most pronounced when accessibility improvements connect regions over long distances, including cross-border trade routes. Urban expansion also correlates with road network density (Zhao et al. 2017), while rural roads remain vital for development, offering safe, affordable, and sustainable transport and local employment (Sharma and Joshi 2020).

Yet roads require regular attention if they are to continue to deliver development benefits. Maintenance is essential to preserve quality, ensure safety, and protect surrounding resources. The Food and Agriculture Organization (FAO 1998) and World Bank (Burningham and Stankevich 2005) warn that neglecting road defects can lead to structural failure, costing up to three times more than regular maintenance. Climate change adds another layer of complexity, with shifting rainfall, temperature, and sea levels affecting road durability (Norwell 2004).

Despite its importance, road maintenance is chronically underfunded. A recent study shows that 10 developing economies (i.e., Bangladesh, Cambodia, Kyrgyz Republic, Mongolia, Nepal, Pakistan, Papua New Guinea, Timor-Leste, Uzbekistan, Vanuatu) allocate only 40% on average, with the funding gap ranging from $8 million to $1.9 billion (Van Dissel and Anyala 2024). One of the main challenges is the collection of appropriate data. Collecting and processing road quality data is expensive, labor-intensive, and often restricted by poor access. Yet without it governments cannot prioritize intervention or prevent costly deterioration.

This data gap also hinders comprehensive assessment of progress toward Sustainable Development Goal (SDG) 9.1, which calls for "quality, reliable, sustainable and resilient infrastructure" with equitable access for all (UN 2023). One key indicator is the Rural Accessibility Index (RAI), which measures the share of the population within 2 km of an all-season road. Studies show that improved rural connectivity, reflected in higher RAI scores, correlates with reduced inequality and poverty (ADB 2002[1]; Sieber and Allen 2018[2]; Alba 2020[3]; Wahyuni et al. 2022[4]). ADB (2022) has highlighted that better rural access fosters innovation, prosperity, and regional stability.

So how do we close the data gap? A number of ADB initiatives offer a glimpse into the future. In the Kyrgyz Republic, road inventory data[5] is being used to train satellite-based algorithms for assessing road condition and vulnerability. In Uzbekistan, projects like the Green Roads Toolkit and Digital Twins for Smart Road Construction integrate remote sensing, sensor data, and simulation platforms to monitor road performance and resilience.

1 This study recognized that roads are a critical enabling condition for improving living conditions in rural areas. However, it also concluded that low-income communities require support to optimize the benefit rural roads confer. On this note, it was suggested that integrated road projects, which include complementary activities to broaden livelihood opportunities, are needed to tackle poverty effectively.

2 The study notes the positive effects of rural roads on income and poverty. This is partly due to increased accessibility to social services and employment. It also emphasized the need to consider the life cycle of roads, given constrained public budgets for road maintenance and poor management. These may shorten the impact of road investments. The study urges greater attention and continued research to better understand how to address poverty and equity with low-carbon transport and increased connectivity.

3 This study determined that RAI scores reflect a nation's urban and rural poverty in relation to infrastructural development. Therefore, RAI can assist national planners and authorities in roadway infrastructure planning.

4 The study calculated RAI results at district level to evaluate the implementation of the Nawacita Program in Indonesia from 2014 to 2020. The program is the main foundation for boosting Indonesia's development, defined as economic growth by prioritizing economic equity and improving government performance (Purnama 2021). The result of the study shows an increase in RAI value and decrease in inequality, specifically in the regions of priority for this program (Papua and West Papua provinces). In view of the impact of rural roads, the study recommended that the government create regulations designating several road infrastructure projects as part of a national strategy. It also identified taxes and private sector investment as possible sources of road infrastructure development funding.

5 The data collection includes measurement and recording of lengths, widths, pavement type, condition, and others; and IRI, rutting, cracking, and others.

Innovative approaches are gaining traction. Meocci et al. (2021) used vehicle black box data to detect road distress very accurately. Brewer et al. (2021) applied transfer learning to predict road quality in Nigeria using United States (US) models, achieving 94% accuracy. Rajput et al. (2022) leveraged global positioning system (GPS) and accelerometer data from public buses to infer road roughness and damage using unsupervised machine learning.

These breakthroughs show that satellite imagery, machine learning, and smartphone-based data can revolutionize road quality monitoring, especially in resource-constrained settings. They offer scalable, cost-effective, and timely alternatives to traditional methods.

This study provides a comprehensive overview of these approaches, covering conventional techniques and contemporary innovations. It complements the *Guidebook on Machine Learning Techniques for Road Quality Monitoring* (March 2025), reviewing traditional data collection methods and exploring geospatial and artificial intelligence (AI)-driven technologies. Special attention is given to compiling the RAI, a critical metric for SDG 9.1. The report is designed for development practitioners and statisticians alike—those who seek to understand, compile, and act on road quality data to drive inclusive and sustainable development.

Compiling Sustainable Development Goal Indicator 9.1.1: Rural Accessibility Index

How close are people to the roads that connect them to socioeconomic opportunities? Answering this question motivates the collection of data for SDG Indicator 9.1.1 - the Rural Accessibility Index. It measures the share of the rural population living within 2 kilometers of an all-season road.

In general, the SDG Global Indicator Framework classifies each indicator into one of three tiers.

(i) Tier 1: The indicator is conceptually clear, has an internationally established methodology and standards available. National data are regularly produced and cover at least half the population in every region where the indicator is relevant.
(ii) Tier 2: The indicator is conceptually clear, has internationally established methodology, and standards, but data are not regularly produced by countries.
(iii) Tier 3: No internationally established methodology or standards are available for the indicator, but a methodology and/or standards are being (or will be) developed or tested.

At the time of writing, RAI is classified under the second tier, meaning we know how to measure it, but many national statistical systems still struggle to do so regularly.

The World Bank, as the official data custodian of SDG 9.1.1, has been working on RAI since 2006 (Iimi et al. 2016), well before the SDGs were even adopted in 2015. In 2006, RAI estimates were arrived at based on relevant information collected from household surveys (e.g., income and expenditure surveys, poverty surveys, core welfare indicators survey, and others) conducted by national statistics offices (Workman and McPherson 2020; Roberts et al. 2006). But these surveys lacked specific questions related to RAI, forcing researchers to interpret survey responses creatively, in the context of estimating RAI scores. This approach led to comparability issues. Lack of spatial representation is a concern too, due to small survey sample sizes (Iimi et al. 2016). Additionally, it was costly to sustain a survey-based approach for the compilation of RAI scores.

To address these technical and administrative challenges surrounding RAI compilation, the World Bank collaborated with the United Kingdom's (UK's) Department for International Development and Research for Community Access Partnership (ReCAP) to refine the methodology. The enhanced methodology proposed the use of geospatial data: the overlaying of rural population maps with road networks to determine accessibility. The enhanced methodology further defined the concept of all-season roads by classifying road conditions based on numbers from the International Roughness Index (IRI) and comparing them with a predefined threshold value.

Rural population data was based on the 1995 Global Rural–Urban Mapping Project (version 1) database. The urban–rural boundaries followed are those set by each national statistics office. The 2019 United Nations Expert Group on Statistical Methodology for Delineating Cities and Rural Areas' DegUrba Methodology will be utilized.

OpenStreetMap (OSM) is the main global data source for road networks. Although rural roads are not as comprehensively depicted in OSM as urban and main roads, there have been improvements given the collaborative nature of OSM.

The 2016 methodology developed by the World Bank used any available source for accessibility factors and road conditions which generally came from official local sources. The resulting definition of an all-season road is as follows:

(i) Paved road with IRI fewer than six meters per kilometer (m/km) and unpaved road with IRI less than 13 m/km, when IRI data are available.
(ii) Paved road in excellent, good, or fair condition and unpaved road in excellent or good condition, where IRI data are not available, but other road condition data, such as the pavement condition index (PCI)[6] or visual assessment by class value, are available."

[6] The Pavement Condition Index (PCI) is a numerical rating that reflects the surface condition of pavement based on visible distress. This method involves visually surveying roads and parking lots to evaluate pavement condition.

The 2016 methodology used by the World Bank to measure geographic information system (GIS)-based RAI is summarized in Figure 3. Road network maps and rural population data are combined to create a map of the proportion of the rural population living more than 2 km from a road. The map also shows those within 2 km of all-season roads by applying road condition data.

Figure 3: Methodology for Measuring the Rural Accessibility Index

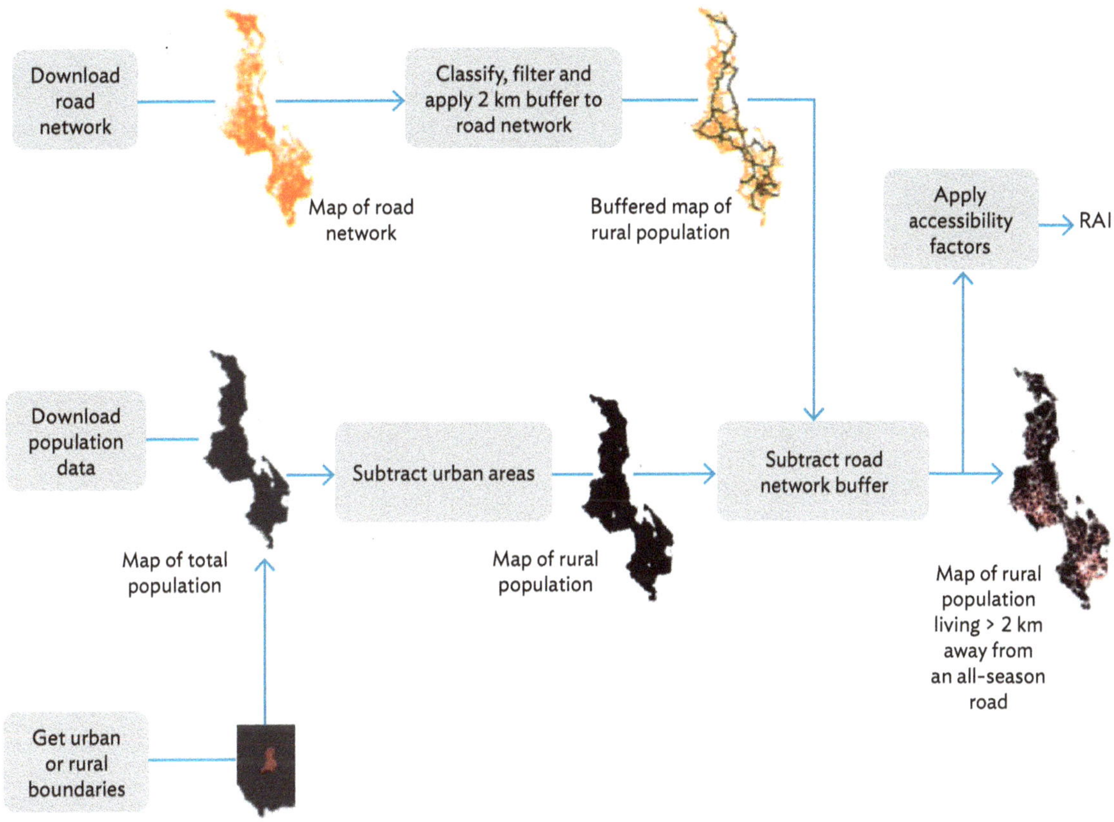

km = kilometer, RAI = rural accessibility index.
Source: R. Workman and K. McPherson. 2020. *Measuring rural access for SDG 9.1.1.* Wokingham, UK: TRL Ltd.

Despite these efforts to standardize relevant concepts and compilation methods, only three economies in Asia and the Pacific—Armenia, Bangladesh, and Nepal—have official RAI estimates (shown in Figure 4) appearing in the UN SDG Global Database at the time of writing.

This data scarcity has prompted efforts to explore alternative compilation methods as discussed in Box 2.

Figure 4: Rural Accessibility Index

Source: Authors based on the United Nations Sustainable Development Solutions Network Sustainable Development Goal Global Database 1 (accessed 15 February 2024).

Box 2: Initiatives on Measuring

The rural accessibility index (RAI) is a key indicator for evaluating rural connectivity to essential services such as healthcare, education, and markets. With advances in geospatial technologies, newer methodologies have emerged, greatly enhancing the precision and practicality of RAI assessment.

The World Bank's 2016 approach to RAI calculation integrated spatial data to analyze the relationship between rural populations and all-weather roads (World Bank Transport and ICT 2016). Building on this, the 2019 Transport Research Laboratory (TRL) Supplemental Guidelines introduced refinements that consider factors beyond road surface quality, leading to a more comprehensive framework (Workman, McPherson, TRL 2019).

A global implementation led by the United Nations Sustainable Development Solutions Network's Sustainable Development Goals (UN SDSN's SDG) Transformation Center uses open datasets and geospatial modeling to standardize RAI measurement across regions, allowing for global comparisons (Iablonovski et al. 2024). Tools like the World Bank's Rural Accessibility Map (RAM), powered by the Open Source Routing Machine, simulate travel times from villages to services, enabling dynamic and localized assessment of rural access.

Studies support and expand on this methodology. For instance, the International Labour Organization's (ILO) 2017 study in Timor-Leste demonstrated how localized road surveys, spatial modeling, and remote sensing can yield more representative RAI estimates, guiding rural investment strategies. It also revealed inconsistencies in the original RAI approach and proposed a cost-effective, spatially detailed alternative (Eqbali, Athmer, and Asare 2017).

A ResearchGate publication recommended spatial information science to capture variations in road conditions and population clusters, particularly in rugged terrain, enabling more equitable rural accessibility evaluation (Akin and Demirel 2019).

The *Transactions in GIS* journal outlined a spatial statistical framework integrating high-resolution road and population datasets, producing disaggregated RAI metrics that enhance planning precision at subnational level (Workman and McPherson 2020).

Source: Authors.

Efforts to improve RAI methodology through the collaboration of the World Bank as the official custodian of SDG 9.1.1, ReCAP, and TRL have resulted in improved data (as discussed in detail in Box 3). Updated RAI data is made available through the RAI measurement tool, although this is in a proof-of-concept stage as of 2019. With these new estimates, a clearer picture of the status of rural access in Asia and the Pacific is emerging.

Box 3: Transport Research Laboratory Supplemental Guidelines

The 2019 Transport Research Laboratory (TRL) Supplemental Guidelines expand on the original World Bank methodology for measuring the Rural Accessibility Index (RAI) by introducing greater flexibility, inclusivity, and practical use of data (Workman, McPherson, TRL 2019). A major advancement in the guidelines is the shift away from evaluating roads solely based on their surface type (such as paved or unpaved), which was central to the 2016 approach. Instead, the TRL guidelines introduce "Accessibility Factors," a broader set of criteria that assess the functional reliability of roads throughout the year. This adjustment is particularly beneficial for low-income countries, where many unpaved rural roads, although lacking formal surfacing, still provide dependable year-round access.

Another important innovation is the integration of open source data guidelines to encourage the use of platforms like OpenStreetMap for mapping road networks and WorldPop for understanding population distribution. By leveraging these openly available, community datasets, governments can conduct RAI measurement more efficiently and cheaply, ensuring assessment is based on current and comprehensive information.

A central theme of the TRL guidelines is emphasis on national ownership and capacity building. They advocate involvement of National Statistical Offices (NSOs) and other key government agencies in the RAI calculation. This approach ensures that data collection and analysis are tailored to specific local contexts and development priorities. Furthermore, involving national agencies promotes accountability, enhances local technical capacity, and ensures the sustainability of RAI monitoring efforts over time.

Recognizing the need for practical tools to implement the updated methodology, a global rollout of the RAI has been initiated using the TRL framework. This effort utilizes global open datasets to provide standardized rural accessibility metrics across territories and regions, allowing for meaningful international comparisons. Additionally, the RAI Measurement Tool, developed by Azavea, offers policymakers and researchers a user-friendly, interactive platform to calculate RAI values at both national and subnational levels. This tool simplifies the complex geospatial analysis behind RAI computation and makes it accessible to a wider range of users.

Overall, the TRL Supplemental Guidelines represent a significant leap forward in rural accessibility measurement. By introducing more flexible, data-driven, and context-sensitive approaches, they significantly enhance the precision, relevance, and utility of the RAI. Through global implementation efforts and the development of digital tools, these guidelines play a crucial role in advancing Sustainable Development Goal (SDG) 9.1.1, helping promote equitable access to infrastructure and essential services for rural populations worldwide.

Source: Authors.

Based on estimates from ReCAP, rural access differs greatly across the region, from less than 25% for Papua New Guinea and Solomon Islands to 100% for Nauru and Singapore. For the whole of Asia and the Pacific, the RAI is 75%, with the highest proportion of the population with rural access coming from the Eastern Asia subregion and high-income economies (ADB 2021). This means that about 560 million people across Asia and the Pacific still live more than 2 km from an all-season road (ADB 2021).

The Asian Transport Observatory (ATO) estimates that the Asia and Pacific RAI will increase to 76% by 2030, and an additional 55 million rural residents will have rural access (Figure 5). With almost 515 million of the rural population still lacking access, the region will be unable to achieve SDG 9.1.1 by 2030.

Figure 5: Estimated Rural Accessibility Index in 2019, by 2030, and Average

(%)

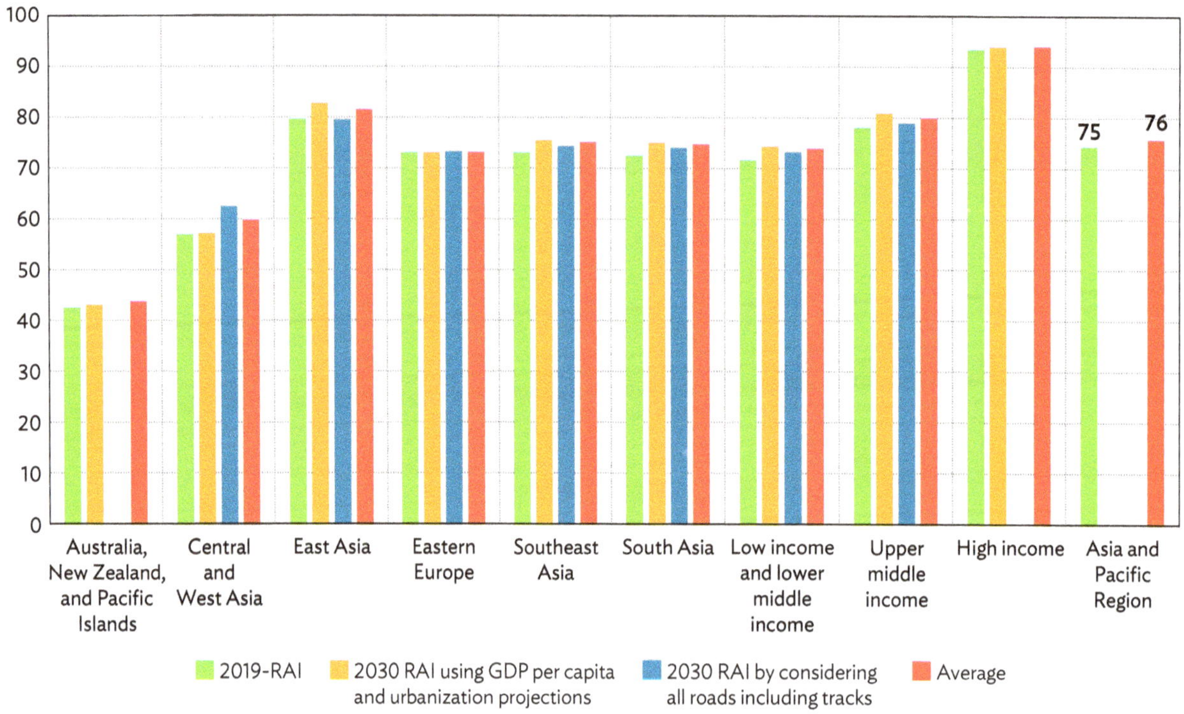

GDP = gross domestic product, RAI = rural accessibility index.

Source: S. Gota, C. Huizenga. 2022. *Asian Transport 2030 Outlook.* https://asiantransportobservatory.org/documents/11/Asian_Transport_2030_Outlook_Nov_2022.pdf.

Following collaboration between the World Bank and ReCAP in 2016 to develop a new methodology for measuring RAI, in 2018 ReCAP launched a new project to continue the improvement and assimilation of RAI methodology. The project aims to "develop a harmonized approach to data collection and measurement of the RAI that is relevant, consistent, and sustainable" (Workman et al. 2020). There are three phases to this project under task group 1 (TG1), task group 2 (TG2), and task group 3 (TG3). TG1,completed in 2018, oversaw an in-depth review and study of the status and comprehensive history of the development of the rural access index.

TG2, building on the output from TG1 and working with World Bank and other stakeholders, sought to enhance and focus the RAI methodology and integrate existing and suggested data collection strategies. The outcome of TG2 is the publication of the RAI methodology Supplemental Guidelines which "clarified data collection, put in place clear quality assurance standards, and provided robust analytical methods to meet international standards" (Workman et al. 2020). As a result, RAI achieved SDG Tier 2 status by December 2018. TG3 set out to "provide [sic] policy guidance for countries measuring RAI, develop [sic] a Roadmap to achieve Tier 1 status by 2025, write [sic] a scientific paper for publication in a relevant journal, and extend [sic] the measurement of RAI that was initiated in TG2" (ReCAP 2020).

Complementing TG2 was the development of the RAI measurement tool which uses open source data from OpenStreetMap, WorldPop (2019), and the Global Rural-Urban Mapping Project in calculating the rural access index as well as the 2016 methodology by the World Bank and the Supplemental Guidelines of TRL. In collaboration with ReCAP and TRL, Azavea developed the software framework for this measurement tool. The proof of concept—found at https://rai.azavea.com/—is an interactive map visualization showcasing rural access index estimates for different economies. The RAI measurement tool also aims to make reporting SDG 9.1.1 much easier by aiding its promotion from Tier 2 to a Tier 1 SDG indicator. It is also to be tested on the UN Global Platform.

Figure 6 shows the RAI estimates for 43 ADB member economies derived from the measurement tool. These data should be considered with caution as they are not yet confirmed by official sources.

Figure 6: Rural Accessibility Index Measurement Tool—Asia and the Pacific
(%)

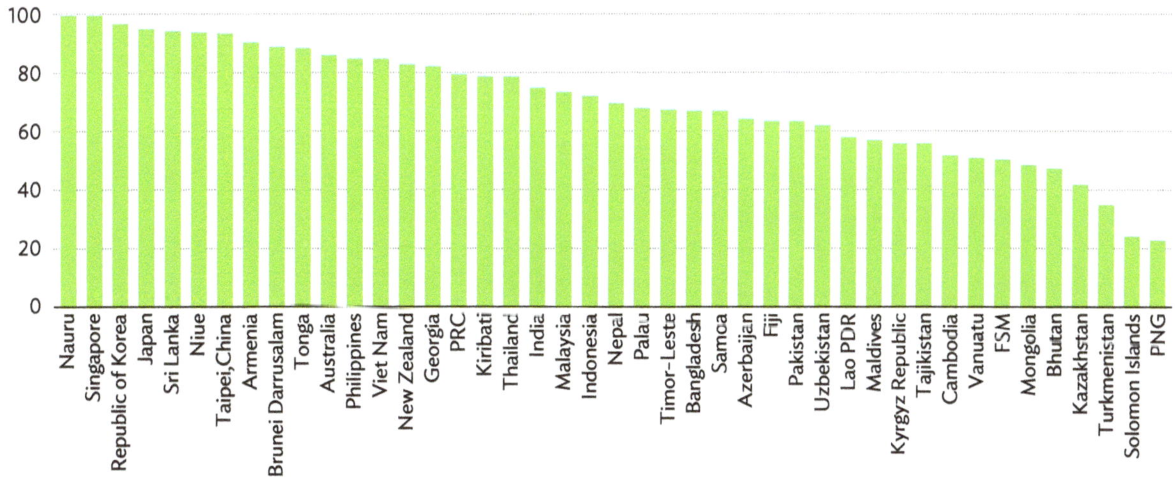

FSM = Federated States of Micronesia, Lao PDR = Lao People's Democratic Republic, PNG = Papua New Guinea, PRC = People's Republic of China.
Source: Rural Access Index Measurement Tool. https://rai.azavea.com/.

The RAI measurement tool was updated in December 2019 and is not being maintained as of 2023. In comparison, the RAI estimates in 2015 for Bangladesh and Nepal, and 2017 for Armenia, found in the UN SDG Global Database, were based on the World Bank February 2019 report. Figure 7 shows a comparison of the RAI from the UN SDG Global Database and the Azavea RAI measurement tool for the three countries with available data. Significant differences in values can be observed comparing the RAI from the two sources. This discrepancy can be explained by the methodology applied for computing the RAI. The World Bank report utilizes the 2016 methodology, which was previously discussed. On the other hand, the Azavea RAI measurement tool considered the 2019 TRL guidelines in the computation. Furthermore, since the RAI data was computed over different years, the available data used for computation may have contributed to the discrepancy.

Figure 7: Rural Accessibility Index (RAI) Comparison—Sustainable Development Goal Global Database Versus Azavea RAI Measurement Tool

(%)

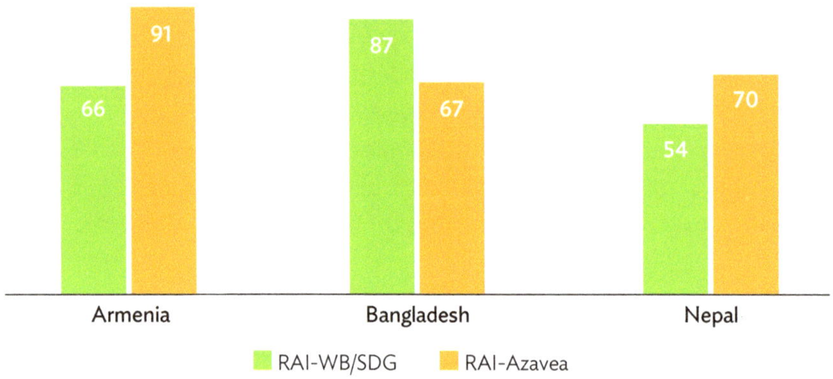

RAI = rural accessibility index, SDG = Sustainable Development Goal, WB = World Bank.

Source: United Nations SDG Global Database. https://unstats.un.org/sdgs/indicators/database/ (accessed 15 February 2024); Rural Access Index Measurement Tool. https://rai.azavea.com/ (accessed 15 February 2024).

ADB launched the Asian Transport Outlook (ATO) to deepen understanding of transport issues in Asia and the Pacific. Covering 51 economies and over 400 indicators, it is a regional knowledge platform developed to guide sustainable transport development in the region. The ATO, (now known as the Asian Transport Observatory) also supports the planning and delivery of transport sector assistance by ADB, as well as transport policy and initiatives by Asian governments, in line with the SDGs, and other international agreements (ATO undated). ATO data is collected from publicly accessible sources comprising primary data from national statistics agencies or transport ministries, and secondary data from other international organizations and think tanks.

The Rural Accessibility Index is one of the indicators included in the ATO. The RAI data presented are procured from three sources: (i) the ReCAP-Azavea RAI Measurement Tool; (ii) the February 2019 World Bank paper *Assessing Rural Accessibility and Rural Roads Investment Needs Using Open Source Data*; and (iii) the Sustainable Mobility for All (SuM4All) Partnership's Rural Access Index Household Survey Methodology (sourced from the World Bank). Although all sources are in the ATO database, the Asian Transport 2030 Outlook report published in November 2022 uses data from the RAI Measurement Tool.

The ReCAP-Azavea RAI Measurement Tool defines rural access as the share of the rural population within 2 km of an all-season road. Its method integrates open datasets such as OpenStreetMap (OSM) for road networks and WorldPop for population distribution, using geospatial analysis to calculate access buffers. It also used the Global Rural-Urban Mapping Project (GRUMP) for urban/rural classification. It is adaptable globally, but was piloted in economies like Kenya and Nepal. Its main limitation is data quality; many roads in rural areas are unmapped or wrongly classified as all-season. It also relies on global rural–urban definitions that may not match national classifications.

The World Bank's 2019 paper follows the same RAI definition, but focuses on estimating rural road investment needs. The methodology relies on remote sensing, open geospatial datasets, such as OSM, WorldPop, Global Human Settlement-Settlement Model (GHS-SMOD). Combining these with assumptions about road conditions to identify rural populations lacking access, an estimate is made of the investment required to improve roads. It has broader geographical modeling potential, but applies generalized assumptions about road conditions and construction costs. Its limitations include limited ground validation and reliance on default road type classifications that may not reflect actual accessibility.

SuM4All's Spatial RAI (SRAI) takes a probabilistic approach, estimating the likelihood that a given area is accessible rather than applying a strict binary (within or outside 2 km of a road). The methodology uses advanced GIS analysis, satellite data, and multilayered proxies such as GHS-SMOD (for urban/rural classification), road classifications from global datasets, and gridded population data. It offers global, subnational estimates across over 180 economies with improved spatial detail. However, it remains limited by data gaps, assumptions about road usability, and the technical complexity of its methods, which require advanced GIS capacity.

All three sources present different methodologies with their limitations, which can be applied to a wide range of economies. However, due to these different methodologies, the resulting RAI data differ as shown in Table 1.

Table 1: Rural Accessibility Index as Reported in the Asian Transport Outlook National Database

Economy	ReCAP-Azavea (2019, %)	World Bank (2019, %)	SUM4ALL
Armenia	90.8	29.5	80.0 (2003)
Australia	86.4		
Azerbaijan	65.0	65.0	67.0 (2002)
Bangladesh	67.4	77.6	87.0 (2015)
Bhutan	47.7	67.6	47.0 (2003)
Brunei Darussalam	89.1		81.0 (1999)
Cambodia	52.2	90.8	81.0 (2003)
People's Republic of China	79.4	71.8	97.0 (2003)
Fiji	63.9		76.0 (1999)
Georgia	82.3	85.4	82.0 (2003)
India	75.2	69.6	61.0 (2001)
Indonesia	72.6	72.6	94.0 (2003)
Japan	95.0		99.0 (2002)
Kazakhstan	42.5	57.0	77.0 (2001)
Korea, Republic of	96.9		89.0 (2003)
Kyrgyz Republic	56.5	64.9	76.0 (1998)
Lao People's Democratic Republic	58.6	70.0	64.0 (2002)
Malaysia	73.6	68.8	82.0 (2001)
Maldives	57.3		
Micronesia, Federated States of	50.6		82.0 (1999)
Mongolia	49.1	42.7	36.0 (2003)
Niue	94.1		
Nauru	100.0		
Nepal	70.0	93.7	54.0 (2015)
New Zealand	82.8		83.0 (2003)
Pakistan	63.8	60.1	61.0 (2004)
Palau	68.1		
Papua New Guinea	23.2		68.0 (1996)
Philippines	85.1	89.5	80.0 (2003)
Samoa	67.2		71.0 (1999)
Singapore	100.0		
Solomon Islands	24.5		77.0 (1999)

continued on next page

Table 1 *continued*

Economy	ReCAP-Azavea (2019, %)	World Bank (2019, %)	SUM4ALL
Sri Lanka	94.4	96.8	92.0 (2003)
Taipei,China	93.6		
Tajikistan	56.4	59.0	74.0 (2003)
Thailand	79.2	93.3	33.0 (2000)
Timor-Leste	67.6		90.0 (2001)
Tonga	88.7		86.0 (1999)
Turkmenistan	35.4	13.2	66.0 (1999)
Uzbekistan	62.5	61.4	57.0 (2000)
Vanuatu			77.0 (1999)
Viet Nam	85.0	88.8	84.0 (2004)

SUM4ALL = Sustainable Mobility for All.
Source: Asian Transport Observatory. https://asiantransportobservatory.org/snd/ (accessed 15 February 2024).

Moving forward, there is a need to strengthen the capacity of statistical systems to ensure the compilation of data required for SDG monitoring, including data for SDG 9.1.1 or the RAI. As mentioned earlier, one of the challenges in compiling RAI datasets is the amount of resources needed to collect data on road quality for a more objective approach to identifying all-season roads (see Chapter 2). The use of remotely acquired data, particularly images taken by satellites or drones and the application of computer vision techniques, could potentially supplement other conventional resource-intensive approaches for road quality monitoring.

Asian Development Bank Technical Assistance

The use of big data presents an exciting opportunity to enhance development statistics, including SDG indicators. New data from satellites, machine-generated sources, social-media, consumer databases, and open sources, can lead to more inclusive, and collaborative problem solving (Sachs 2012). To continue working successfully toward sustainable development, it is critical to embrace a data revolution where traditional data sources (surveys, censuses, and administrative databases) are augmented with innovative new sources. An impeccable coalescence of data collected by national statistical systems and other data producers from both public and private institutions is essential for a successful data revolution (ADB 2016).

To address the limited availability of data on road quality metrics such as roughness of road, used to compile SDG indicator 9.1.1 on RAI, it is instructive to explore the feasibility of using big data. ADB statisticians are collaborating with development partners to pilot geospatial data, particularly satellite imagery and computer vision techniques, to generate road quality data for areas where such information is not available using conventional approaches. A machine learning-based methodology for predicting the roughness of roads may have wider applications beyond catering to data requirements for monitoring SDG target 9.1. Such methodology could be considered a preliminary assessment tool in identifying which roads need to be rehabilitated, especially if such infrastructure is in remote areas and less likely to be monitored regularly. The methodology may help ADB developing member country (DMC) transport agencies reduce monitoring costs and enable them to identify the most effective preventive maintenance strategy.

This study is part of a wider ADB knowledge and support technical assistance (TA) project: *Using Frontier Technology and Big Data Analytics for Smart Infrastructure Facility Planning and Monitoring*. It is funded by the Government of Japan through the ADB administered Japan Fund for Prosperous and Resilient Asia and the Pacific Technical Assistance. The project aims to identify, demonstrate, and evaluate the use of high-level technologies in the effective planning of high-quality infrastructure investments focusing on transportation, and measuring the economic impact. This project seeks to improve the physical and financial sustainability of transport infrastructure in DMCs by utilizing available and emerging technology and tools. Spatial information, artificial intelligence, machine learning, and the internet of things are some examples that can be harnessed to improve the planning, design, construction, and maintenance of infrastructure. The pilots and case studies in this TA will help DMC governments facing the challenge of identifying the appropriate technology option.

The TA hopes to improve transport and infrastructure monitoring and maintenance in ADB DMCs. The first component of the project is a feasibility study that could help develop a road map for capitalizing on innovative technologies and data sources for infrastructure quality monitoring. Conventionally collected data on road quality are used to develop machine learning algorithms to predict transport infrastructure quality and/or other development indicators. This technical report focuses on the first component of the project, using data from select economies in developing Asia, particularly the Philippines and Thailand.

Developing Asia has several initiatives aimed at distributing investment in infrastructure to address deficits and sustain growth. These initiatives encompass a range of projects that target specific areas of transportation infrastructure, including the construction or improvement of roads, bridges, ports, transport terminals, flood control infrastructure, and communication facilities. As these projects expand access to roads, it is critical to consider cost-effective methods for monitoring the condition of the infrastructure.

II Review of Approaches for Road Pavement Condition Assessment

Road Performance: Structural and Functional Performance

Roads are designed and maintained to ensure their primary objectives are achieved. These are firstly, for the road paving to have the structural capacity to cope with vehicle weight, and secondly, to ensure passenger comfort along with satisfactory skid resistance. These aspects of pavement performance are evaluated with the information obtained from various pavement condition assessments and are used by highway engineers to determine how well a pavement is performing.

Various types of condition assessments are required for this purpose. The data for pavement condition assessments are obtained from pavement condition surveys.

Pavement condition is an important aspect of road quality maintenance decision-making. The need to quickly identify pavement deterioration is critically important to road agencies. If a road agency does not make timely repairs to paved roads in the early stage of deterioration, the condition may rapidly worsen, and maintenance costs increase significantly. The use of pavement condition data to support maintenance and resurfacing becomes more critical as data-driven approaches increase.

Both manual data collection (observation on site) and automated data collection (that use line and area scanners, ground-penetrating radar, acoustic sensors, optical imagery, LIDAR, and others) are used for this purpose. Use of these techniques depends on an agency's priorities, available resources, frequency of the survey required, and geographic limitations. The type of evaluation carried out with the results of the condition assessments will also dictate the type of techniques used for the survey. Two challenges in pavement management are the timely detection of problems and the frequency of data collection. Many road agency studies show that the timely detection of problems and treatment with preventive measures increase the service life of the road, reduce total maintenance costs, and maintain safety and quality.

Performance Metrics for Road Quality Assessment

Pavement Roughness

Pavement roughness is a nondestructive evaluation parameter that gives an indication of the ride quality and is considered a primary functional condition evaluation parameter. The measurement of road roughness is important for the management of economic road maintenance. Roughness is an important pavement characteristic because it affects not only ride quality, but also vehicle delay costs, fuel consumption, and maintenance costs (Greene et al. 2013; World Bank Publications, 1987). Because of the importance of roughness for road asset management decision-making, road agencies devote considerable resources trying to measure road roughness in a repeatable and reproducible manner.

Pavement roughness is generally defined as an indicator of unevenness in a pavement surface. The unevenness and resulting vibration on the vehicle adversely affect the ride quality of the user. The roughness is an indication of cumulative vertical displacement recorded due to the irregularities on the pavement over a measured distance. This cumulative measure of ups and downs in road profile is termed as the roughness index or unevenness index and is normally expressed in m/km or mm/km. Roughness also refers to the longitudinal profile of a pavement (Figure 8).

Figure 8: Pavement Roughness Derivation from a Longitudinal Profile

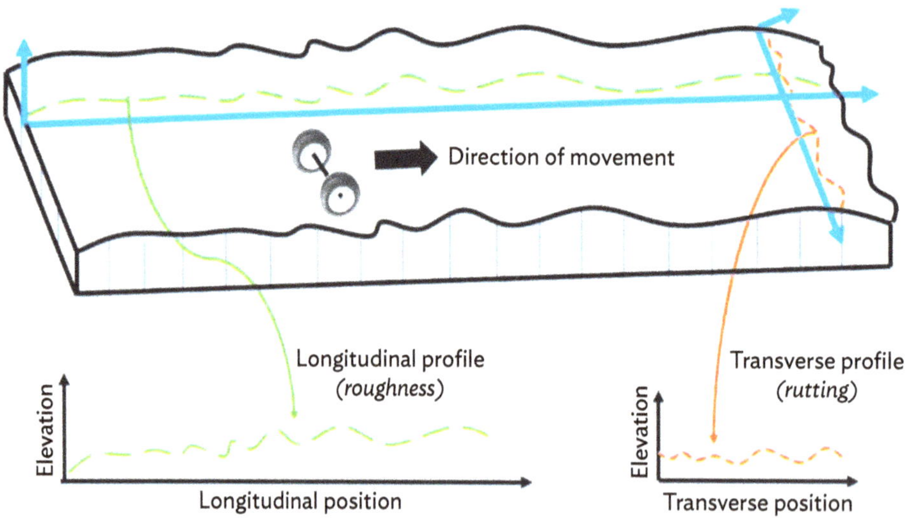

Source: K. Gkyrtis, A. Loizos, and C. Plati. 2021. "Integrating Pavement Sensing Data for Pavement Condition Evaluation" Sensors 21, no. 9: 3104. https://doi.org/10.3390/s21093104.

Pavement roughness can be measured by using various indexes such as the ride number (RN), the IRI, the half-car roughness index (HRI), the Mays ride meter (MRM), the quarter-car index (QI), and the present serviceability index (PSI). Table 2 provides the definition and measurement methodology for each index.

Table 2: Pavement Roughness Indexes

Index	Definition	Measurement
RN	A subjective smoothness rating from 0 (poor) to 5 (excellent), reflecting user perception.	Calculated via regression between profilometer-measured profile index (PI) and Mean Panel Rating (MPR).
IRI	A standardized roughness metric (m/km or in/mi) representing the accumulated suspension movement per distance traveled, derived from a "quarter-car" vehicle model simulating ride response.	Obtained by profiling the road surface (e.g. inertial profilers), feeding the data into the quarter-car model, then calculating the average absolute suspension displacement divided by distance.
HRI	A variant of IRI that uses a "half-car" model simulating a two-axle vehicle, typically computed from averaged left and right wheel path profiles.	A trailer-mounted spring–damper measures vertical motion caused by pavement irregularities; the result is expressed as an index (e.g., inches per mile).
MRM	A mechanical roughness index based on the vertical response of a sprung mass in a trailer-mounted system.	A trailer-mounted spring–damper measures vertical motion caused by pavement irregularities; the result is expressed as an index (e.g., inches per mile) .
QI	The result of applying a "quarter-car" dynamic model, i.e., a simplified suspension system representing one wheel, to pavement profile data.	Process a single wheel path profile through the quarter-car model; output is used in IRI computation. Essentially, QI is the core underlying roughness calculation in IRI.
PSI	A composite rating from 0 (failed) to 5 (excellent) covering pavement smoothness, rutting, cracking, and patching; a refined surrogate for user-perceived serviceability.	Calculated via AASHTO's empirical formula using quantifiable metrics like slope variance (SV), rut depth, and cracking/patching extent . PSI is also strongly correlated with IRI.

HRI = half-car roughness index, IRI = international roughness index, m/km = meter per kilometer, MRM = Mays ride meter, PSI = present serviceability index, QI = quarter-car index, RN = ride number.
Source: Authors.

Among those, IRI is globally accepted as a suitable parameter to measure pavement roughness. The International Road Roughness Experiment held in Brazil in 1982 developed and adopted the IRI in m/km as the unit to define roughness. Table 3 shows the range of IRI values for different pavements and the corresponding speeds. (Abdelaziz et al. 2020).

Table 4 is an illustration of the American Standard of Testing and Materials (ASTM) standard E1926–98, and the Standard Practice for Computing International Roughness Index of Roads from Longitudinal Profile Measurements.

Table 3: Comfortable Speed for Different Roughness Values

IRI Value (m/km)	Comfortable Speed (km/h)	Description
2	Over 120	Very smooth
4	100–120	Smooth, with a few surface imperfections
6	70–90	Perceptible movement, frequent distresses such a minor depression
8	50–60	Some swaying and wheel bounce, frequent distresses with moderate severity with minor potholes, raveling and others.
10	40–50	Significant swaying, frequent distresses, with moderate to high severity, include potholes, rutting, and raveling
12	30–40	Consistently rough, severe distresses, with major depressions, extensive areas of potholes present
14	< 30	Very rough surface hard to drive safely on

IRI = international roughness index, km/h = kilometers per hour, m/km = meters per kilometer.
Note: American Standard of Testing and Materials (ASTM) standards.
Source: Authors.

Acceptable roughness values can also be established based on the operational speed of the road, as the resulting vibration, and others induced by the pavement roughness would be affected to an extent by the speed of the vehicle (Table 4) (Yu et al. 2006). This is an important consideration for road agencies, as the definition of maintenance thresholds can vary depending on the type or hierarchy of the road within the network. For example, to initiate rehabilitation works, an expressway with higher operational speeds (90–100 km/h) would have a lower maximum acceptable IRI value (e.g., 3–4 m/km) compared to the IRI threshold (8–10 m/km) of a low volume rural road which would have lower operating speeds (30–40 km/h).

Table 4: International Roughness Index Thresholds at Different Speeds and Ride Quality Levels

Ride Quality Level	International Roughness Index Thresholds at Different Speeds (m/km)					
	20	40	60	80	100	120
Very Good	<5.72	<2.86	<1.90	<1.43	<1.14	<0.95
Good	5.72–8.99	2.86–4.49	1.90–2.99	1.43–2.24	1.14–1.79	0.95–1.49
Fair	9.00–11.39	4.50–5.69	3.00–3.79	2.25–2.84	1.80–2.27	1.50–1.89
Mediocre	11.4–16.16	5.70–8.08	3.80–5.40	2.85–4.05	2.28–3.24	1.90–2.70
Poor	>16.16	>8.08	>5.40	>4.05	>3.24	>2.70

m/km = meters per kilometer.
Source: J. Yu et al. 2006. Development of speed related ride quality thresholds using international roughness index. Transport Research Record.

The relationship for equivalent IRI values corresponding to different speeds is given as follows (Múčka 2017).

$$IRI\,(v_2) = \left(\frac{v_1}{v_2}\right)^{0.5} IRI\,(v_1)$$

Where this choice of IRI values should meet the condition of the same quarter-car model suspension relative velocity response for two velocities. For example, for two road speed limits, v_1 = 50 km/h (local roads) and v_2 = 110 km/h (highways), IRI (110)/IRI (50) = 0.67 should hold.

Overview of Pavement Condition Assessment Methods

Roughness

The roughness measurement devices are also categorized by their accuracy and the data collection frequency. The ASTM standard E 950 (ASTM International, 2004) defines roughness into four categories from Class 1 to Class 4 based on the precision level, sampling interval, bias, and others. Class 1 and 2 measurements measure the actual road profile with high precision. Class 3 measurements are based on the response type and roughness estimation based on the correlation while Class 4 is categorized as a subjective rating and uncalibrated methods (Table 5).

Table 5: Summary of Each Class and Respective Measurement Accuracy

Class	Longitudinal Sampling	Vertical Measurement Resolution	Precision (1 SD) (mm)	Bias (mm)
Class 1	Less than or equal to 25 mm	Less than or equal to 0.1 mm	0.38	1.25
Class 2	Greater than 25 mm	Greater than 0.1 mm	0.76	2.50
Class 3	Greater than 150 mm	Greater than 0.2 mm	2.50	6.25
Class 4	Greater than 300 mm	Greater than 0.5 mm	–	–

mm = millimeter; SD = standard deviation.
Source: ASTM International. 2004. *Standard Test Method for Measuring the Longitudinal Profile of Traveled Surfaces with an Accelerometer Established Inertial Profiling Reference: E 950–98*. ASTM International.

Pavement Roughness Evaluation Methods

Table 6 shows the classification of various devices in practice with the classes of IRI measurement, based on IRI measurement accuracy and precision level.

Table 6: Classification of Devices Used to Measure the International Roughness Index

Class	Description	Specification	Commercial Examples	Applications
Class 1	• The highest standard of accuracy for measurement of IRI • Requires the measure of accurate elevation points closely spaced along the traveled wheel path • Class 1 device is viewed as a primary method to validate other methods	• Measurement error of less than 0.3 m/km for paved roads and of 0.5 m/km for all other road types; maximum longitudinal sampling intervals ≤ 250 mm • Precision of vertical elevation measures 0.5–3 mm	Noncontact laser profilers, face dipstick, ARRB walking profiler, rod and level	Most detailed and comprehensive (network and project level applications)
Class 2	• Profile measurement with accuracy less than that required in Class 1 due to random and bias errors over a range of conditions • Includes IRI determined from profiles measured by high-speed profilometers or a static device which do not satisfy the required sampling intervals and precision	• Measurement error of 0.3–0.5 m/km for all other road types; maximum longitudinal sampling intervals ≤250 mm and ≤500 mm • Precision of vertical elevation measures of 1–6 mm	Profilometers (e.g., California, Rainhart); General Motors Research (GMR) inertial profilers	Detailed (network and project level applications)
Class 3	• IRI estimated from correlation equations using response type road roughness measurement systems (RTRRMS) • Such devices typically determine roughness using accumulated suspension motion to measure average rectified slope (ABS) or accelerometers as transducer • RTRRMS must be calibrated by correlation to be considered Class 3 method	• Road condition rated by class (i.e., good, fair, and poor) or numerically with a measurement error of 0.5–1 m/km	Road master; Road Measurement Data Acquisition System (RODMAS); rolling straight edge	Summary (network and project level applications)
Class 4	• Subjective evaluation of road condition including the user's perception of ride quality and visual inspection, or using measurements from uncalibrated instruments	• Road condition rated by class (i.e., good, fair, and poor) or numerically with a measurement error of 2–6 m/km	Visual inspection	Most summary

ARRB = Australian Road Research Board, GMR = General Motors Research, IRI = international roughness index, m/km = meters per kilometer, mm = millimeter, RODMAS = road measurement data acquisition system, RTRRMS = response type road roughness measurement systems.
Source: Authors.

Class 1 Measuring Equipment: Precision Profilers

Precision profiles have the highest standard of accuracy for calculating the IRI and consist of a series of accurately measured elevation points, closely spaced along the segment.

A rod and level survey is the simplest way of measuring roughness to provide an accurate measurement of the pavement profile. Due to the instrument being immobile during measurement, this is known as a static method. Profiling devices are faster than the rod and level method, which include battery onboard computers to automatically record data and compute roughness. This type of profiler has an error of less than 0.3–0.5 m/km for all road types while maximum longitudinal sampling intervals are ≤250 mm. Moreover, precision levels of the vertical elevation measures are 0.5–3.0 mm.

These devices are profiled along the line of walking and include a precision inclinometer, which measures the height difference between two points. The use of surveys for large projects, however, is impractical due to time and cost. These devices are commonly used for calibration purposes.

Laser profilers are high-speed inertial profilers. They are used to minimize errors and standardize the survey process by recording, reducing, processing, and storing pavement data using equipment attached to a multifunction vehicle (Australian Road Research Board 2021).

Class 2 Measuring Equipment: Other Profilometric Method (High-Speed Inertial Profilers)

In the 1960s, the General Motors Research Laboratory developed the first high-speed inertial profiler to use network measurements to monitor large road networks. The profilers are combined with three measurements to evaluate roughness (i.e., a reference elevation, a height relative to the reference, and longitudinal distance).[7] With these types of roughness measurement techniques, an accelerometer is used to convert vertical acceleration to instant heights using a data processing algorithm. The height of the ground relative to the reference elevation is measured by roughness measurement equipment, which is the distance between the accelerometer in the vehicle and the ground directly beneath (Figure 9). These profiler measurements are less accurate than that required in Class 1 due to random and bias errors over a range of conditions. This type of roughness measurement error is also 0.3–0.5 m/km; however, the maximum longitudinal sampling interval is less than 250 mm and ≤500 mm. Moreover, the precision of vertical elevation measures is 1–6 mm (Spangler and Kelly 1966).

This height is measured with a noncontacting sensor such as a laser, infrared or ultrasonic sensor. A vehicle speedometer or odometer is used to measure the longitudinal distance in these types of instruments.

[7] Reference Elevation: The height above or below a fixed reference point. Height Relative to the reference: The height difference between the ground and measuring unit. Longitudinal Distance: The distance travelled by the roughness measuring equipment.

Figure 9: A Typical High-Speed Inertial Profiler

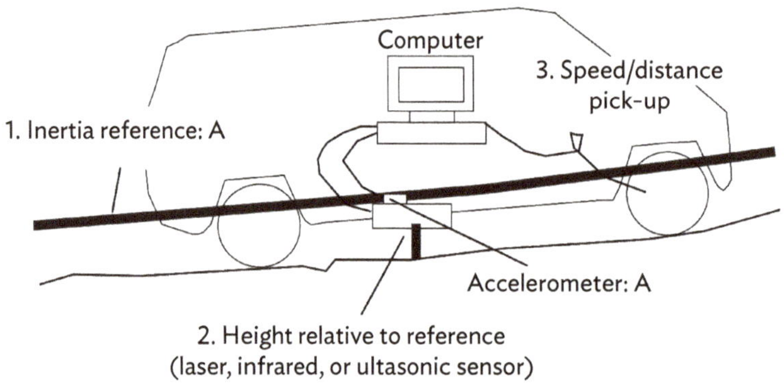

Source: M. W. Sayers, S. M. Karamihas. 1998. *The Little Book of Profiling: Basic Information about Measuring and Interpreting Road Profiles.* The Regent of the University of Michigan.

Class 3 Measuring Equipment: Response Type and Correlation-Based Method

This category is known as response-type road roughness measuring systems (RTRRMS), response-type systems, or road meter systems. In these systems, the measurement is taken as accumulated suspension strokes over a traveled distance reported in counts per mile (counts/mi) and then converted into IRI m/km or inches per mile (in/mi). The measure of vehicle response is similar to the acceleration of the vehicle body and, therefore, highly correlated with ride vibration. The arrangement of the response type systems is shown in Figure 10. Due to portability and

Figure 10: Arrangement of a Response Type Roughness Measurement System

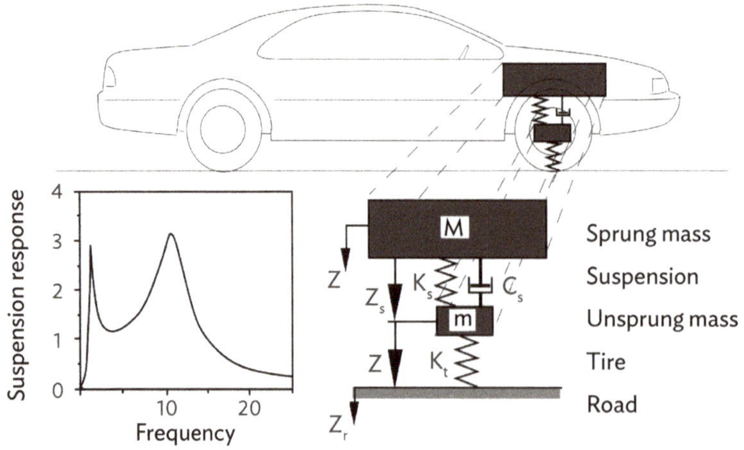

C_s = damping of the uncontrolled suspension, K_s = stiffness of the uncontrolled suspension, K_t = tire stiffness, m = unsprung mass, M = sprung mass, Z_r = road displacement, Z_s = sprung mass displacement, Z = unsprung mass displacement.
Source: M. W. Sayers, S. M. Karamihas. 1998. *The Little Book of Profiling: Basic Information about Measuring and Interpreting Road Profiles.* The Regent of the University of Michigan.

reproducibility, these measurements are commonly used in 2023. The Roughometer, Mays meter, Bump Integrator, MERLIN, and Riley are types of equipment used for response type measurements and correlate the response into roughness measurements using a defined computer algorithm.

Class 4 Subjective Rating and Panel Rating

Long before high-speed profilers, engineers used this method to gauge the opinion of road travelers. Normally, these would be people driving on the road section in question and asking them to rate the pavement. Human ratings are inevitably subjective and time bound, compared with objective scientific roughness rating data.

The present serviceability rating (PSR) is based on an engineer's judgment as to the current capability of a pavement to appropriately allow vehicles to use it. It is known as a ride quality measurement such as roughness and ranges from 5 (excellent) to 0 (very poor). The PSR is known as the basis for IRI categorization. Table 7 shows the PSR description based on Federal Highway Administration (FHWA) interpretation as well as a threshold from the highway performance monitoring system field manual (Al-Omari and Darter 1994).

Table 7: Present Serviceability Rating Description per Federal Highway Administration Guidelines from HPMS Field Manual

Present Serviceability Rating 5 (excellent) **to 0** (very poor)	**Description**	**Rating** (Highway Performance Monitoring System Field Manual)
5.0–4.0	Only new (or nearly new) superior pavements are likely to be smooth and distress-free enough (sufficiently free of cracks and patches) to qualify for this category. Most roads constructed or resurfaced during the data year would be rated in this category.	Good
4.0–3.0	Pavements in this category, although not quite as smooth as those categorized above, give a first-class ride and exhibit few, if any, visible signs of surface deterioration. Flexible pavements may be beginning to show evidence of rutting and fine random cracks. Rigid pavements may be beginning to show evidence of slight surface deterioration, such as minor cracks and spalling.	Fair
3.0–2.0	The ride qualities of roads in this category are noticeably inferior to those of new pavements and may be barely tolerable for high-speed traffic. Surface defects of flexible pavements may include rutting, map cracking, and extensive patching. Rigid pavements in this group may have a few joint failures, faulting and/or cracking, and some pumping.	

continued on next page

Table 7 *continued*

Present Serviceability Rating 5 (excellent) **to 0** (very poor)	**Description**	**Rating** (Highway Performance Monitoring System Field Manual)
2.0–1.0	Pavements in this category have deteriorated to such an extent that they affect the speed of free-flow traffic. Flexible pavements may have large potholes and deep cracks. Distress includes raveling, cracking, and rutting and occurs in over 50% of the surface. Rigid pavement distress includes joint spalling,[a] patching, slab cracking,[b] and scaling[c] and may include pumping and faulting.[d]	Poor
1.0–0.0	Pavements in this category are in an extremely poor condition. The facility is passable only at reduced speeds and with considerable ride discomfort. Large potholes and deep cracks exist. Distress occurs over 75% or more of the surface.	

Notes:

[a] Joint Spalling: Loss of a section of concrete pavement from the surface or along the edges of cracks and joints.

[b] Slab cracking: Transverse cracks may appear parallel to joints and can be caused by thermal stresses, poor subgrade support, or heavy loading.

[c] Scaling: Surface deterioration that causes loss of fine aggregate and mortar.

[d] Pumping and faulting: Joints and cracks may become faulty or develop a step between adjacent slabs.

Source: *Highway Performance Monitoring System (HPMS) Field Manual.* https://www.fhwa.dot.gov/policyinformation/hpms/fieldmanual/hpms_field_manual_dec2016.pdf.

Mean panel rating, ride number, and profile index are other common subjective rating systems that have been defined in the past. Through the findings of several research studies, a relationship was developed to predict the IRI based on the PSR (Equation 2).

The typical relationship between the IRI and the PSR is represented in an exponential form in the equation below.

$$PSR = \alpha IRI^{-\beta * IRI} \tag{2}$$

Where **α** and **β** are the constants to be predicted based on the road agency categorization and IRI in m/km. The major issue in these measures is the inconsistency and dependency on the person's perspective, which is not suitable for the network level.

Pavement Condition Data Required for Condition Evaluation

The type of pavement distress data that is required for network-level[8] or project-level[9] condition evaluation is defined by several factors. For example, rural local road pavement condition data would differ from an urban expressway network. While some core issues, such as distress data, would be relevant in either case, there will be more emphasis on certain types of condition data based on the functionality of the road within the road hierarchy.

For network evaluation, it may not be necessary to obtain data on all the inconsistencies in the road. Most of the time, what is required is the overall condition of the pavement that would be an outcome of the distresses present. On some occasions, details of distresses such as potholes, rutting, and alligator cracking may be collected as they could be used to assess the structural performance of the pavement. Project evaluation would generally include a more detailed evaluation which may cover all aspects of structural and functional performance. Therefore, more types of data are necessary to carry out a more accurate evaluation.

Pavement evaluations at network level are not done very frequently. Also, road networks under road agencies include road of several thousands of km, meaning data collection is very time-consuming. If it is for donor funding decisions, such projects typically have a multiyear implementation period. Even if maintenance is done as part of recurrent expenditure from regular budget allocations, it would take at least 1–2 years for the project to be executed from the time condition data is monitored, analyzed and the maintenance needs evaluated and for the maintenance works to be executed. Network level monitoring of pavement condition data may be required around the 2–3-year interval. Project data collection would be required to be carried out rapidly as and when necessary.

Data richness is the opportunity provided by data to gain useful knowledge that can improve decision-making. This refers to the level of detail required for the given assessment metric. For example, if the overall pavement condition is considered as the assessment metric, is a numerical value representing the condition required or would a categorical analysis suffice? The richness of the data required for the evaluation would be a key consideration in determining the type of monitoring tool that would be required to obtain it. This will determine the time and cost of data collection and the frequency at which it can be monitored if implemented. The accuracy and precision of the data are also related to this. At the project level, data richness is a key requirement since the accuracy of the costing and evaluation of remedial measures would depend on accurate and precise data with a sufficient level of detail.

8 Network-level condition evaluation focuses on the entire road network.
9 Project-level condition evaluation focuses on specific road sections or corridors identified for intervention.

Network Evaluation

Network pavement condition evaluation serves several objectives for road agencies.

(i) It facilitates the prioritization of road segments for maintenance or rehabilitation activities or allows preliminary screening to be carried out to identify candidate road links for further investigation;
(ii) It allows preliminary estimation of the budget requirement for road improvements that will be planned; and
(iii) It provides input for geographic information systems to visualize overall network condition.

Network-level pavement condition data are typically collected in larger quantities and at relatively higher survey speeds. The automation of data collection is necessary since the data has to be collected over a shorter period. The data collected by road agencies using sensors (e.g., roughness, rut depth, faulting) can be estimated with higher accuracy and precision (McGhee 2004).

The type of pavement condition data collected at the network level differs from those for projects. For example, the IRI, friction (skid number), rut depth, faulting, and surface distress data are collected at the network level, but structural evaluation related parameters are collected primarily at the project level (Flintsch and McGhee 2009). The pavement condition data typically collected at network level evaluation is given in Table 8.

Table 8: Typical Data Types Collected at the Network Level Evaluation

Country/State/Road Agency	Types of Roads in the Network	Pavement Condition Evaluation Metrics	Pavement Condition Data Collected and the Measurement Parameters
Japan, Kazakhstan, New Zealand, UK, Alabama, Austroad, British Columbia, Columbia, Cyprus Maryland, Oklahoma, New Jersey	Urban	Roughness	IRI
Malaysia	Trunk roads		BI unit
Japan, Kazakhstan, Malaysia, New Zealand, UK, Alabama, British Columbia, Columbia, Maryland, Oklahoma, New Jersey, Austroad	Urban	Surface Distresses	Rut depth
Japan, Kazakhstan, Malaysia, UK, Maryland, New Jersey	Urban		Cracking
Alabama, Oklahoma	Urban		Faulting
Kazakhstan, Malaysia	Urban		Raveling
Kazakhstan	Urban		Patching
Cyprus	All road		Major deterioration % area
Kazakhstan, Malaysia, Austroad	Trunk roads	Structural capacity	Deflection
New Zealand, UK, Austroad	State highway	Pavement Texture	MPD (mean profile depth)

continued on next page

Table 8 *continued*

Country/State/Road Agency	Types of Roads in the Network	Pavement Condition Evaluation Metrics	Pavement Condition Data Collected and the Measurement Parameters
UK	National highways	Skid resistance	Skid number

BI unit = bending index unit, IRI = international roughness index, MPD = mean profile depth, UK = United Kingdom.
Sources: Pierce and Zimmerman, 2014; Michael Baker International, 2018; Pantuso et al. 2019; Snaith, 1998; Kubo, 2017; New Zealand Transport Agency, 2021; Moffatt, Sharp, and Ferguson, 2006; Department of Transport, London, 2021; Onn, 2001.

Project Evaluation

Project evaluation is often required while maintenance work is being implemented. The evaluation should be done such that the right maintenance treatment is identified with a more accurate cost estimate. This is particularly relevant for identifying structural issues. Project evaluation is also carried out, for example, during performance-based contracts where the agency is evaluating the performance metrics prescribed in the contract to which the developer must conform. When there are public complaints or accident investigations, detailed analysis, for example, of the skid resistance of the segment would be carried out.

Project evaluation would require more detailed pavement condition data, with several additional characteristics for the particular pavement condition monitoring parameter included (that may define severity and extent), as well as an increase in accuracy and precision of the measurement. The road segment length that needs to be covered would be relatively low compared to the network evaluation.

Traditional Techniques of Pavement Condition Assessment

Given the importance of monitoring road pavement conditions, there are several methods and techniques for data collection. These methods can be divided into two categories: manual/visual techniques and automated data collection technologies. Manual/visual techniques refer to pavement condition assessments that require human involvement on-site to record the characteristics of the pavement being assessed. Automated data collection is when pavement data collection is carried out with the assistance of sensors, video, images, scanners, radar, laser, and others.

The data is still collected in the field and the equipment required to acquire the data via these methods is normally attached to a survey vehicle on the road. The second type of data collection requires minimal human input and utilizes technology. It also requires additional analytics to transform the data into parameters used to assess pavement condition. Table 9 outlines details of some automated data collection technologies.

Table 9: Automated Data Collection Technologies

Method/Technique	Description
Laser profiling	Laser profilers are used to minimize errors and standardize the survey process by recording, reducing, processing, and storing pavement data using equipment attached to a multifunction vehicle. In this technique, a network survey vehicle is equipped with a digital laser profiler which can estimate various data types with higher accuracy and precise levels. Most of the equipment can collect data for a 3.6 m wide road, but it reduces survey time and costs due to its ability to collect data at highway speed (usually up to 100 km/h). This type of profiler has advantages over other techniques as results are independent of vehicle type, measurements are possible on all sealed surfaces, and data can be linked to Chainage and GPS coordinates.
Light Detection and Ranging (LiDAR)	LiDAR is a remote sensing method that uses pulsed lasers to measure variable distances to the Earth. These light pulses are combined with other data from airborne system to generate precise, three-dimensional information regarding the shape of the Earth's surface (NOS 2021).
Video-image processing	Video or image processing techniques involve a camera that leverages computer algorithms or neural network methods to evaluate pavement conditions, such as potholes, cracking, aging, depressions, texture, and others.
Ground Penetrating Radar (GPR)	GPR is an electromagnetic nondestructive geophysical tool that is widely used for pavement investigation, including the differentiation of pavement layers, the mapping of layer thicknesses, and the assessment of pavement condition (Aleksey et al. 2017). GPR is predominantly used to determine pavement thickness and subsurface defects in most situations, while some researchers have evaluated the effectiveness of GPR to measure pavement surface distress.
Vehicle response measurement methods	Vehicle response measurement methods are largely applicable to roughness detection. Class 3 roughness measurement equipment such as the bump-integrator has been used to measure roughness along with the GPS coordinates of the road section. Roughness is measured by using vehicle body vibration over a measured distance using odometers.
Thermal mapping	Thermal mapping is a vehicle survey technique that identifies and quantifies the distribution of temperature differences across a highway network (VAISALA 2021). The spatial variation of road surface temperature is measured using a high-resolution infrared thermometer. Then it identifies patterns of temperature variation, by accurate measurement. Synchronized thermographics are developed to evaluate the pavement condition as well as the levels of distress.

GPR = ground penetrating radar, GPS = global positioning system, km/h = kilometer per hour, m = meter.
Source: Authors.

Box 4 provides a brief discussion of how these methods are used in select developing economies: the Philippines and Thailand.

Box 4: Methods in Measuring the International Roughness Index in the Philippines and Thailand

Philippines

The Department of Public Works and Highways is the government agency mandated to design, construct, and maintain national roads, bridges, and flood control. They accumulate and store road quality databases in hard copies and manual entries. A pavement management system is used to assist the agency in decision-making for maintenance. In terms of road quality monitoring, they collect IRI data and present serviceability rating data through the use of laser profilers. Ideally, road quality is monitored annually. Technical difficulties with equipment are commonly encountered while surveying. In response, the agency has purchased the latest equipment to improve road quality monitoring. The agency also has experience in using several alternative technologies in road quality monitoring, such as smartphone applications, satellite imagery, drone images, and SAR. These technologies have resulted in some advantages, including the ability to assess roads that are not easily accessible, centralized monitoring, and the availability of historical data. They also have disadvantages, such as potential inaccuracy in road quality monitoring, a requirement for technical resources, and extensive processing time.

Thailand

The Department of Highways is the government agency responsible for managing major roads. This includes construction, maintenance, and operation, among others. They accumulate and store road databases using MS Excel and cloud storage. They also offer road quality databases in hard copies from manual entries. A pavement management system is used to assist the agency in maintenance decision-making. They consider IRI for road quality metrics, which are ideally collected every 2.5 years, although the agency was only able to collect IRI data every 4–5 years. Technical difficulties with equipment are also common issues encountered during the conduct of the survey. In response, the agency purchases the latest equipment to improve the efficiency of road quality monitoring and conduct capacity building for the use of new tools and techniques. The agency also has experience in using smartphones to collect road quality data. This technology was able to assess the condition of roads that are not easily accessible, but require a lot of time to assess.

IRI = international roughness index, SAR = synthetic aperture radar.
Sources: "Survey on the Use of Frontier Technology and Big Data for Enhanced Infrastructure Management" conducted by the Authors in 2022.

Use of Satellite Imagery and Machine Learning for Road Quality Monitoring

Conventional road quality monitoring is generally an expensive undertaking. Data on road quality, such as road roughness, are routinely collected only for select areas (e.g., primary highways in urban areas). To extend data collection to a wider area, such as remote rural regions, it is imperative to explore more cost-effective approaches.

There is literature on measuring road quality using machine-learning algorithms, particularly neural networks trained on large datasets. In 2008, Gerke and Heipke used the sequential nature of roadway networks to improve the extraction of roadway imagery from satellite imagery. They suggest that a network or sequential-based approach may be useful for measuring road quality by utilizing the proximity of road segments.

Mnih and Hinton (2010) developed a neural network to detect roads from high resolution aerial imagery. It uses a procedure that makes use of Restricted Boltzmann machines road map data which specifies the centerline of each road. Their results were superior to existing published research based on automatic road detection.

From 2018 to 2022, other studies were produced about road quality monitoring and machine learning. In 2018, Cadamuro et al. developed a convolutional neural network (CNN) to predict road quality in Kenya. The study used IRI data as a measure of road quality, covered 1,150 km of road, and considered five classes of road quality based on IRI data and a binary classification model for the prediction. The resulting model achieved an accuracy rate of 64% for five classes and at least 87% on the binary classification.

Oshri et al. (2018) used a residual neural network with transfer learning to correctly classify road quality in African countries. They combined infrastructure assessments from Afrobarometer data with satellite imagery. With an accuracy rate of 70.5%, the model was trained separately in urban and rural areas, as well as across the countries. However, the authors noted their concerns about potential overfitting.

On a 157 km motorway in Portugal, Marcelino et al. (2018) applied a regularized regression with a lasso algorithm to predict an indicator for road quality. They used continuous data such as IRI, mean profile depth, and the cracking area. The results showed models from the machine learning algorithm had lower mean squared errors than conventional methods of examining pavement conditions. The study concluded that machine learning techniques are a promising solution for predicting road quality relative to conventional road evaluation.

In predicting road surface distress, Eisenbach et al. (2019) analyzed nearly 2,500 images of road surfaces from Germany. They trained three neural network machine learning models (VGG, ResNets, and Support Vector Machine) on a binary decision problem (distress vs. no distress). Regarding prediction accuracy, residual neural and VGG networks outperformed more conventional machine learning models like support vector machines. ResNet was found to be the most suitable for detecting pavement distress since it uses less computing power.

Abdelaziz et al. (2020) determined that standard CNN resulted in an R squared of 0.75, outperforming a linear regression classification model of 0.57. They analyzed long-term pavement performance data from the United States government to predict IRI classification on highway data from the United States.

Fiorentini et al. (2021) conducted a study to determine the robustness of machine learning algorithm (MLA) predictions of road quality as a viable alternative to profilometric road surveys. The study applied a persistent scatter interferometric algorithm to obtain data from satellite images and compared these to IRI values from a laser profiler. The study determined that the significant correlations between conventional IRI values and MLA predictions were due to external factors. Weak correlations were attributed to local factors, such as traffic loading. The authors recommend calibrating MLAs with local road conditions to improve prediction accuracy.

Bashar and Torres-Machi (2021) applied artificial neural network (ANN), random forest (RF), and support vector machine (SVM) algorithms to predict IRI scores using data from a few selected studies. Random forests showed the best results in predicting IRI scores with an overall performance of 0.995, while ANN had consistently accurate results over a significant number of studies. SVM algorithms showed different results from ANN and random forest models.

Leduc and Assaf (2022) applied the AutoML model and CNN to identify and classify road damage using 298 road images captured using GoPro camera at an angle of 60°. The study showed that the AutoML model identified damage with precision ranging from 0.50 to 0.92. Visual data used to validate the AutoML results showed 71% true positives.

Table 10 summarizes the aforementioned studies.

Table 10: Studies Using Satellite Imagery and Machine Learning for Road Quality Monitoring

Study	Measure	ML Methodology	Data Sources	Coverage	Key Findings/ Outputs
Gerke and Heipke (2008)	Conformity of vector road maps with actual roads observed from images	Graph-based reasoning combining extracted image features with vector data, classical image processing and statistical modeling	Aerial/satellite imagery + vector data	Germany (2007–2008)	Semi-automated validation ~66% accuracy
Mnih and Hinton (2010)	Detect roads from high resolution aerial imagery	CNN	Aerial imagery (US urban)	US urban (~2009)	High detection accuracy
Cadamuro et al. (2018)	IRI (binary and 5-class classification)	CNN	50 cm satellite + IRI (Kenya)	Kenya (~2015)	88% binary, 73% multiclass accuracy

continued on next page

Table 10 *continued*

Study	Measure	ML Methodology	Data Sources	Coverage	Key Findings/ Outputs
Oshri et al. (2018)	Road potholes and defects detection	CNN	Satellite imagery + survey/Africa	Africa (2015–2018)	Automated defect detection validated
Marcelino et al. (2018)	Pavement distress metrics (cracking, roughness, rutting, skid resistance), summarized via composite Performance Indicator	Statistical aggregation; limited machine learning	Brazil pavements and inspections	Brazil (2016–2017)	Developed multimetric performance indicator
Eisenbach et al. (2019)	Pavement defect detection from visual imagery	CNN object detection	Vehicle camera images	Europe (2019)	DL greatly improved inspection automation
Abdelaziz et al. (2020)	IRI for flexible pavements	ANN, decision trees	Structural and traffic data (Egypt)	Egypt (2017–2019)	R^2 > 0.8 predictive accuracy
Fiorentini et al. (2021)	IRI via PS-InSAR and ML	ML+PS-InSAR	LiDAR, PS-InSAR, profilometer (Italy)	Italy (2020)	RMSE ~0.6 m/km vs. profilometer
Bashar and Torres-Machi (2021)	IRI prediction	RF, boosting, SVM regression	Spanish PPM datasets (Spain)	Spain (2015–2019)	MAE ~0.9 m/km from ML models
Leduc and Assaf (2022)	Urban road quality visualization	CNN + graph modeling	IoT, urban cameras, sensor networks	European smart-city	Real-time high-resolution quality maps

CNN = convolutional neural network, DL = deep learning, IRI = international roughness index, LIDAR = light detection and ranging, m/km = meter per kilometer, MAE = mean absolute error, ML = machine learning, PPM = pavement performance modeling, RF = random forest, RMSE = root mean squared error, SVM = support vector machine, US = United States.
Source: Authors.

Use of Computer Vision Algorithms for Predicting Road Quality

Neural Networks

Neural networks are a computer model inspired by how the human brain works. Just like our brains learn from experience, neural networks can learn to do tasks by studying examples. They are especially good at solving complex problems, like recognizing faces in photos, translating languages, or predicting stock market trends.

A neural network is made up of layers of nodes, or tiny decision-makers, which are interconnected. These connections allow information to pass through the network, get processed, and produce an answer, like recognizing a photo as a cat or a number.

To train a neural network, a large amount of labeled data is required. For example, if we want it to recognize handwritten numbers, we show it thousands of pictures of digits along with the correct answer. Over time, the network learns to spot patterns in the images, like curves or lines, and uses those to make accurate guesses.

A typical neural network has three parts. First, the input layer receives the raw data, like the pixel values of an image. Next are the hidden layers, which process the data by spotting features or patterns. Last, the output layer makes the final prediction or classification.

For example, to recognize the number 3 in a small 8x8 image, the network looks at all 64 pixel values. It passes these through layers that detect basic shapes (like lines or curves), then combines this information to decide which number it sees.

Convolutional Neural Network

One special type of neural network, a convolutional neural network (CNN), is designed to handle images more efficiently. Instead of looking at the entire image all at once, CNNs scan it using small filters, similar to looking through a tiny window that slides across an image. These filters detect simple features like edges or corners, and deeper layers combine them to recognize more complex shapes, like eyes, wheels, or letters.

CNNs are very good at working with images because they reuse the same filters across the image to make them faster. They also focus only on important parts to help reduce noise, and they do not need people to manually define what to look for.

This makes CNNs extremely useful in things like facial recognition, traffic sign detection, and satellite image analysis.

Using Satellite Imagery for Road Quality Monitoring

Satellite imagery is a cheap and efficient alternative for preliminary road quality assessment, particularly where on-site surveys are not feasible. Given the regularity of satellite image acquisition, this method allows for more consistent and scalable monitoring of road conditions over time. Its implementation is also less labor-intensive, making it especially beneficial for countries short of people trained in road infrastructure monitoring. Publicly available satellite data can provide reasonable levels of detail, allowing assessment at a resolution of up to 10 meters (e.g., Sentinel-2), and can help streamline early identification of road sections requiring maintenance.

Despite these advantages, satellite assessment comes with notable limitations. One significant challenge is the inability to identify specific types of pavement distress or subsurface issues without further on-site investigation. Technical constraints in the imagery itself also affect reliability. For instance, synthetic aperture radar (SAR) imagery is subject to geometrical distortions such as foreshortening and decorrelation, due to changes in terrain and weather, atmospheric interference, and speckle noise, all of which complicate interpretation.

Hyperspectral imagery, while rich in detail, may suffer from low resolution, spectral mixing, and the presence of artifacts, since one pixel may represent various items on the ground. Additionally, inconsistencies in ground truth data used to train and validate machine learning models, such as variations in road segment length and assessment methodology, further complicate analysis. These limitations are particularly prevalent when using free or open-source imagery, although purchasing high-resolution satellite data remains an option. This approach can be expensive, though.

In practice, satellite road condition monitoring is best viewed as a complementary tool rather than a replacement for traditional on-site surveys. It is particularly effective in identifying road sections in good condition and can classify roads into broad categories (e.g., good, satisfactory, poor) with reasonable accuracy. This supports the prioritization of maintenance and can reduce the number of unnecessary site inspections. Nonetheless, applying satellite methods to fully replace condition metrics, like pavement roughness or detailed distress mapping, at scale remains limited due to the high cost and the computing power required to process large, high-resolution datasets.

The effectiveness of a satellite monitoring system depends heavily on selecting the appropriate imagery. Freely available medium-resolution imagery may be suitable for simple classification, though issues such as cloud cover and image timeliness must be considered. More advanced monitoring, especially where detailed assessments are needed, requires high-resolution imagery. Although costly, ranging from approximately $11-$40 per sq km depending on coverage, these images can be acquired more affordably with cost sharing among government departments with overlapping data needs.

From a technical standpoint, the computing requirements for processing satellite imagery are manageable for most road agencies. A standard setup with a 12-core processor, 16 GB of RAM, and a GPU with 4 GB of dedicated memory is sufficient for training machine learning models such as U-Net and random forests. But successful deployment also relies on the capacity of agency personnel. Agencies must decide whether to manage data processing internally or outsource it to specialists, similar to existing pavement monitoring approaches. Regardless of the path chosen, training and support for internal staff are essential. Furthermore, any machine learning models used must be trained and validated using local data to reflect regional variations in road types, terrain, and climate. This helps ensure the accuracy and applicability of the model outputs.

Adopting Smartphone-Based Assessment Methods

The increasing spread of smartphones offers a promising avenue for assessing pavement conditions. Most modern smartphones are equipped with sensitive 3- or 6-axis accelerometers capable of detecting road roughness and defects through vibration from a vehicle's suspension and tires. These sensors, originally designed to enhance user experience of the phone, can now be harnessed for infrastructure monitoring. Smartphone methods offer several benefits: ease of installation; low initial and operational costs; compatibility with multiple vehicles with minimal setup; and potential for large-scale data collection. This especially true if the technology is made available to the public. Limitations include the relatively low correlation between smartphone-generated data and established pavement roughness indicators. There is also a higher risk of data quality issues if the system is publicly distributed without sufficient training or safeguards.

From a technical standpoint, smartphone assessment systems are considerably easier to adopt than traditional monitoring tools that require specialized equipment. Installation involves ensuring the phone is oriented correctly inside the vehicle, with the screen facing upward and its front aligned with the vehicle's direction of travel. The smartphone's x, y, and z axes can then capture motion data corresponding to left/right, front/rear, and up/own directions, respectively. Supplementary devices such as dashboard GPS units and video cameras can also be used to enhance data collection. In terms of cost, this approach is more affordable than conventional methods. Low purchase and operating costs make it accessible for various stakeholders and have led to its successful implementation in road maintenance decision-making in several locations.

Adopting smartphone-based systems at scale does comes with notable limitations. Challenges arise in data collection, information management, and software development, especially when trying to capture individual instances of pavement distress. For broader assessments like road section roughness, some of these issues are less impactful. A key technical challenge is the inconsistency in sensor signals, as repeated passes over the same road anomaly may not yield uniform readings. Sensor behavior can vary significantly depending on the smartphone model, its hardware configuration, and even its operating system. Moreover, uploading large volumes of sensor data can lead to network congestion, causing delays or data loss unless the application is optimized for minimal data usage. Scalability becomes an issue if the system is expanded to crowdsourced public users, placing additional strain on backend information technology (IT) infrastructure.

GPS limitations also emerge, particularly in rural or remote areas where weak satellite signals may result in data gaps. While this is more problematic for detecting specific distress locations, it is less of a concern when measuring general road segment conditions. Privacy concerns must also be considered, as apps that track a user's location to pinpoint road anomalies might raise data security issues. Furthermore, disparities in the quality and sensitivity of smartphone sensors, including accelerometers and GPS chips, can affect the precision of the data collected. Driver behavior further complicates data consistency. For example, a cautious driver may swerve or slow down before hitting a pothole, compromising the app's ability to detect that anomaly accurately.

User engagement is another challenge. The success of a crowdsourced data collection effort hinges on adequate participation, which in turn requires effective incentives. If users perceive no direct benefit, or if the app appears intrusive or complex, the volume and quality of data may be insufficient. Additionally, users may assume that by simply using the app, issues like potholes will be addressed promptly by the authorities. Managing these expectations is crucial, and implementing systems to provide feedback or updates to users could improve participation and trust.

Looking ahead, several improvements are being explored to enhance the effectiveness of smartphone-based road monitoring. Although numerous algorithms have been proposed to detect road surface anomalies using sensor data, comparing their performance remains difficult due to a lack of publicly available models and datasets. Many existing approaches rely on unclear or arbitrary threshold settings, while machine learning methods require extensive training data. A more robust solution would be a hybrid system capable of learning from real-time sensor inputs and contextual geographical data to continuously and accurately identify various road anomalies.

For such a system to work optimally, sensor data must be smoothed and reoriented to accommodate diverse user behavior and increase detection precision. Ideally, a free, cross-platform mobile application would be developed that is self-adapting, self-learning, and fully automatic, requiring no user interaction and allowing drivers to remain focused on the road. To ensure practicality, the app's power consumption, particularly from GPS use, should be minimized. Advanced sensor calibration techniques should be employed to reduce noise and device variability. Another option to explore is establishing a partnership with major IT platforms, such as Google Maps or Gaode Map (Amap). Their advanced technology, proficiency in GIS, and, most notably, their large user communities could provide effective ways to overcome these limitations.

The growing number of smartphone users and the integration of wireless technologies like Bluetooth, wi-fi, and cellular networks make mobile devices ideal for large-scale crowdsourcing of road condition data. Additionally, emerging vehicle communication technologies, such as vehicle-to-vehicle (V2V) and cellular vehicle-to-vehicle (C-V2V), have great potential for improving transport management and road safety. These technologies could allow vehicles to continuously update and share real-time road surface condition data and notify other drivers of upcoming anomalies. However, these innovations remain under development and are not yet widely available, due to the lack of supportive road infrastructure. As such, while smartphone-based assessment methods are already proving useful, their full potential will be realized only with continued technological advancement, careful system design, and strategic integration into broader transportation frameworks.

Viability of Using Both Traditional and Frontier Technology in Road Quality Monitoring

The path forward in modernizing road condition assessment lies in integrating frontier technologies into asset management systems. By leveraging the unique strengths of various tools, road agencies can enhance monitoring processes and optimize maintenance, planning or rehabilitation projects. Each technology plays a distinct role in assessing road conditions, particularly in terms of the metrics it can effectively measure. Current evidence suggests that these advanced technologies are most effective at evaluating overall road surface conditions at the network level, rather than detecting individual types of distress. Key metrics for monitoring pavement condition include both qualitative and quantitative ratings, with pavement roughness emerging as a widely used and reliable parameter.

Satellite-based image processing has proven particularly useful for initial screening, offering a cost-effective way to classify roads into broad condition categories. This method is especially advantageous for road networks that cover large geographical areas, where traditional on-site assessment would be costly and require lots of staff. Roads identified as being in good condition after the initial screening can typically be excluded from further analysis. For roads classified as satisfactory or poor, further evaluation using more detailed and quantitative data becomes essential. Pavement roughness, which can be accurately assessed using smartphone methods, is appropriate for this purpose. Research confirms that these techniques provide sufficient precision for maintenance decisions to be made while remaining cheap and easy to deploy.

Once roughness data are integrated into the asset management system, road agencies can pinpoint which sections, if any, need additional information to make a decision on intervention. This can be collected through manual surveys or image-processing technologies capable of identifying specific types of pavement distress. Manual methods are low tech, but labor-intensive and time-consuming. Automated image-based approaches make data collection speedy, but require greater investment in equipment, trained personnel, and logistical support. Subjected to such detailed evaluation, using earlier screening and smartphone methods narrows the focus of these more resource-intensive assessments.

Strengthening Statistical Capacity on Integrating Big Data into Road Accessibility and Quality Monitoring Information Systems

This study introduces both traditional and emerging methods for assessing road conditions, addressing the persistent gap in road quality data across many economies. While conventional techniques, such as laser profilers, remain widely used, as employed by the Department of Highways (DOH) of Thailand and the DPWH of the Philippines, their application still faces challenges. According to a survey conducted in 2022 by ADB, technical issues have led to the need to replace some equipment, leading to delays in data collection. The lack of access to remote areas further compounds the difficulty of maintaining timely and comprehensive road assessment programs.

Emerging technologies offer alternative approaches that may help address these limitations and enable more frequent and extensive monitoring. However, implementing these innovations introduces its own set of challenges.

Access to Innovative Technologies

Based on the aforementioned 2022 survey, the DOH and DPWH have begun exploring alternative data sources, including smartphone applications, satellite imagery, drone photography, and SAR. These technologies have enabled agencies to collect data from otherwise inaccessible areas, establish centralized monitoring systems, and leverage historical data for longitudinal analysis.

Despite these advantages, several drawbacks remain. First, converting raw outputs into usable road quality metrics often involves time-consuming processes. Secondly, concerns about data accuracy and resolution have been raised, particularly by the Philippine agency, along with the need for advanced technical capabilities and computing infrastructure. These issues highlight the importance of balancing innovation with practical feasibility when adopting new tools for infrastructure monitoring.

Privacy Issues and Concerns Around Big Data

While big data presents significant opportunities to enhance road quality monitoring, such as through mobile sensors, geotagged images, and crowdsourced reports, it also introduces serious privacy risks.

The UN Global Pulse (2012) emphasized the wide implications of privacy concerns in big data, particularly in areas of data acquisition, storage, use, and dissemination. Users of digital services or sensors may be unaware that their data is being collected, shared, or analyzed. Moreover, studies have shown that even anonymized datasets can be vulnerable when cross-referenced with other information sources, raising risks of identification and surveillance.

Additionally, the increasing resolution of satellite imagery and advances in image analysis tools have heightened the possibility of inadvertently identifying individuals or sensitive locations. As noted by Santos (2019), this creates a complex set of ethical, legal, and data protection concerns. These need to be addressed before fully integrating such tools into public infrastructure monitoring systems.

Institutional Strengthening

In 2023, a series of regional workshops held in Japan, Thailand, and the Philippines brought together transport, statistics, and planning staff, as well as academics, from 12 economies across Asia and the Pacific. These workshops examined methodologies and findings from a feasibility study on the use of remotely sensed data (e.g., satellite imagery, surface temperature, and terrain gradients), paired with artificial intelligence, to evaluate road conditions.

A key objective was to promote the adoption of innovative data sources and machine learning techniques for transport infrastructure analysis. Participants expressed a strong interest in applying AI, particularly convolutional neural networks (CNNs) and other deep learning models, for predictive modeling and road condition classification.

Notably, there was a growing demand for hands-on, skills-based training, particularly in using programming tools like Python to develop and apply machine learning models. Participants emphasized the need for continued institutional capacity-building in areas such as AI, remote sensing, and geospatial analysis.

To support this shift, partnerships with commercial technology providers can be explored to facilitate access to high-resolution satellite data and provide training resources. Additionally, the development and sharing of best practice, case studies, and legal guidelines, particularly concerning data privacy, can help ensure the responsible and effective integration of these technologies.

References

Abdelaziz, N. et al. 2020. International Roughness Index prediction model for flexible pavements. *International Journal of Pavement Engineering*. 21(1). pp. 88–99. https://doi.org/10.1080/10298436.2018.1441414.

Adler, M. et al. 2020. *Roads, market access and regional economic development*. OECD Regional Development Papers, No. 6, OECD Publishing, Paris, https://doi.org/10.1787/8b9eca17-en.

Al-Omari, B. and M. Darter. 1994. Relationships between international roughness index and present serviceability rating. *Transportation Research Record Journal of the Transportation Research Board*. pp. 130–136.

Alba, C, 2020. Is the Rural-Access-Index (RAI) a Reflection of Rural and Urban Poverty? A Data Analysis. *International Journal of Social Science and Business*. 4(4). pp. 422–428.

Aleksey, K. K. et al. 2017. Utilization of air-launched ground penetrating radar (GPR) for pavement condition assessment. *Construction and Building Materials*. pp. 130–139.

Asian Development Bank (ADB). 2002. *Impact of Rural Roads on Poverty Reduction: A Case Study-Based Analysis.*

ADB. 2009. *Infrastructure for a Seamless Asia*. http://hdl.handle.net/11540/187.

ADB. 2017. *Meeting Asia's Infrastructure Needs.* Manila. https://dx.doi.org/10.22617/FLS168388-2.

ADB. 2022. ATO. Undated. Asian Transport Outlook National Database (accessed 15 February 2024). https://asiantransportobservatory.org/snd/.

ADB. 2021. *Mapping the Spatial Distribution of Poverty Using Satellite Imagery in the Philippines.* https://dx.doi.org/10.22617/TCS210076-2.

ADB. 2022. *Strategies for Recovery: COVID-19 and Urban Transport Policy in Asia.*

ATO. 2025. *Asia and the Pacific's Transport Infrastructure and Investment Outlook 2035.* https://asiantransportobservatory.org/analytical-outputs/asia-transport-infrastructure-investment-needs/.

ASTM International. 2004. *Standard Test Method for Measuring the Longitudinal Profile of Traveled Surfaces with an Accelerometer Established Inertial Profiling Reference: E 950-98*. ASTM International.

Australian Road Research Board (ARRB). 2021. ARRB Systems. *Hawkeye 2000 Series*. https://www.arrb.com.au/hawkeye-2000.

Bashar, M. Z. and C. Torres-Machi. 2021. Performance of Machine Learning Algorithms in Predicting the Pavement International Roughness Index. *Transportation Research Record*. 2675(5). https://doi.org/10.1177/0361198120986171.

Brewer, E. et al. 2021. Predicting road quality using high resolution satellite imagery: A transfer learning approach. *PLoS ONE*. 16(7). e0253370.

Burningham, S. and N. Stankevich. 2005. *Why Road Maintenance is Important and How to Get it Done*. World Bank Publications—Reports 11779. The World Bank Group.

Cadamuro, G. et al. 2018. Assigning a Grade: Accurate Measurement of Road Quality Using Satellite Imagery. *arXiv*. https://doi.org/10.48550/arXiv.1812.01699.

Department of Transport, London. 2021. *Technical Note: Road Condition and Maintenance Data*. London: Department of Transport.

Eisenbach, M. et al. 2019. *Enhancing the Quality of Visual Road Condition Assessment by Deep Learning*. World Road Congress.

Fan, S. and C. Chan-Kang. 2005. *Road Development, Economic Growth, and Poverty Reduction in China*. Research Report of the International Food Policy Research Institute. https://www.researchgate.net/publication/5056962_Road_Development_Economic_Growth_and_Poverty_Reduction_in_China.

Florentini, N. et al. 2021. Can Machine Learning and PS-InSAR Reliably Stand in for Road Profilometric Surveys? *Sensors*. 21(10). p. 3377. https://doi.org/10.3390/s21103377.

Flintsch, G. W. and K. K. McGhee. 2009. *Quality Management of Pavement Condition Data Collection*. Transport Research Board.

Food and Agriculture Organization of the United Nations. (n.d.). Watershed Management Field Manual: Road Design and Construction in Sensitive Watersheds. *FAO Conservation Guide*. 13/5. Ch. 7 Road Maintenance.

Gerke, V. and C. Heipke. 2008. Image-based quality assessment of road databases. *International Journal of Geographical Information Science*. 22(8). pp. 871–894. https://doi.org/10.1080/13658810701703258.

Gkyrtis, K., A. Loizos, and C. Plati. 2021. "Integrating Pavement Sensing Data for Pavement Condition Evaluation" Sensors 21, no. 9: 3104. https://doi.org/10.3390/s21093104.

Gota, S. and C. Huizenga. 2022. *Asian Transport 2030 Outlook*. https://asiantransportobservatory.org/documents/11/Asian_Transport_2030_Outlook_Nov_2022.pdf.

Government of Japan, Japan International Cooperation Agency. 2014. *Data Collection Survey on Infrastructure Management in Thailand*. Kokusai Kogyo Co., Ltd.:Tokyo.

Government of New Zealand, Transport Agency. 2021. *National pavement conditions*. Wellington. https://www.nzta.govt.nz/roads-and-rail/highways-information-portal/tools/national-pavement-conditions/.

Government of Philippines, Department of Public Works and Highways. 2011. *Minimum Pavement Thickness and Width of National Roads*. Department Order No. 22, Series of 2011.

Government of the United States, National Ocean Service (NOS). 2021. *What is lidar?* National Oceanic and Atmospheric Administration. Washington: Department of Commerce. https://oceanservice.noaa.gov/facts/lidar.html.

Greene, S. et al. 2013. *Pavement Roughness and Fuel Consumption*. Concrete Sustainability Hub.

Iimi, A. et al. 2016. New Rural Access Index: Main Determinants and Correlation to Poverty. *Transport and ICT Global Practice*.

Kasraian, D. et al. 2019. *The impact of urban proximity, transport accessibility and policy on urban growth: A longitudinal analysis over five years*.

Kubo, K. 2017. *Pavement Maintenance in Japan*. Road Conference 2017 International Symposium.

Leduc, E. and G. J. Assaf. 2022. Road visualization for smart city: Solution review with road quality qualification. *Internet of Things*. 12(100305). https://doi.org/10.1016/j.iot.2020.100305.

Marcelino, P. et al. 2018. Comprehensive performance indicators for road pavement condition assessment. *Structure and Infrastructure Engineering*. 14(11). pp. 1433–1445. https://doi.org/10.1080/15732479.2018.1446179.

McGhee, K. H. 2004. *Automated Pavement Distress Collection Techniques*. Transport Research Board.

Meocci, M. et al. 2021. An innovative approach for highperformance road pavement monitoring using black box. *Journal of Civil Structural Health Monitoring*. 11. pp. 485–506.

Michael Baker International. 2018. *Pavement Data Collection Data Quality Management Plan*. New Jersey: Michael Baker International.

Mnih, V. and G. E. Hinton. 2010. Learning to Detect Roads in High-Resolution Aerial Images. In K. Daniilidis et al, eds. *Computer Vision – ECCV 2010*. ECCV 2010. Lecture Notes in Computer Science. pp. 210–223. Berlin, Heidelberg: Springer. https://doi.org/10.1007/978-3-642-15567-3_16.

Moffatt, M. A. et al. 2006. *Austroads Standardised Measurement of Road Condition*. ARRB Conference – Research into Practice. Canberra, Australia.

Múčka, P. 2017. International Roughness Index specifications around the world, Road Materials and Pavement Design. *Road Materials and Pavement Design*. pp. 929–965.

Norwell, G. 2004. *Impact of Climate Change on Road Infrastructure*. Austroads Publication No. AP–R243/04.

Onn, L. K. 2001. *Pavement Data Collection Stratergy for JKR Malaysia*. ARRB Transport Research Ltd Conference. Melborne, Australia:ARRB.

Oshri, B. et al. 2018. *Infrastructure quality assessment in Africa using satellite imagery and deep learning*. 24th ACM SIGKDD International Conference on Knowledge Discovery and Data Mining (KDD '18). pp. 616–625. New York, NY, USA:Association for Computing Machinery.

Pantuso, A. et al. 2019. Analysis of Pavement Condition Survey Data for Effective Implementation of a Network Level Pavement Management Program for Kazakhstan. *Sustainability*.

Pierce, L. M. and K. A. Zimmerman. 2014. *Quality Management for Pavement Condition Data Collection*. Applied Pavement Technology, Inc.

Purnama, S. and P. Bestari. 2020. *Implementation of National Medium Term Development Plan (RPJMN) of the Nawacita Program in Improving Good Governance*. Journal of International Conference Proceedings 3(4).

Rajput, P. et al. 2020. *Pervasive and Mobile Computing*. Elsevier B.V.

Research for Community Access Partnership. 2020. *Rural Access Index (RAI)*.

Roberts, P. et al. 2006. *Rural Access Index: A Key Development Indicator*. Washington, D.C.: The World Bank Group.

Sachs, J. 2012. From Millennium Development Goals to Sustainable Development Goals. *The Lancet*. 379. pp. 2206–2211.

Sayers, M. W. and S. M. Karamihas. 1998. *The Little Book of Profiling: Basic Information about Measuring and Interpreting Road Profiles*. The Regent of the University of Michigan.

Setboonsarng, S. 2005. Transport Infrastructure and Poverty Reduction. *ADBI Research Policy Brief No. 21*. Asian Development Bank Institute.

Sharma, A. and. S. Joshi. 2020. *Rural Roads are Key to Helping Society's Most Vulnerable*. Manila: ADB https://blogs.adb.org/blog/rural-roads-are-key-helping-societys-most-vulnerable.

Siebert, N. and H. Allen. 2018. Impacts of Rural Roads on Poverty and Equity. *Transport and Communications Bulletin for Asia and the Pacific, No.86.*

Smith, M. N. 2016. *The number of cars worldwide is set to double by 2040*. World Economic Forum. https://www.weforum.org/stories/2016/04/the-number-of-cars-worldwide- is-set-to-double-by-2040.

Snaith, S. M. 1998. *The development and implementation of pavement management systems: a case study over 15 years*. Proceedings of the Institution of Civil Engineers—Transport. pp. 72–79.

United Nations. 2023. *Goal 9: Build resilient infrastructure, promote sustainable industrialization and foster innovation*. Sustainable Development Goals. https://www.un.org/sustainabledevelopment/infrastructure-industrialization/.

VAISALA. 2021. *Thermal Mapping*. https://www.vaisala.com/en/product/1381.

Van Dissel, S. C. M. Anyala. 2024. Road Maintenance Financing and Cost Recovery Options: The Future of Road User Revenues in Developing Asia and the Pacific. ADB Sustainable Development Working Paper Series. http://dx.doi.org/10.22617/WPS240549-2.

Wahyuni, R. et al. 2022. Inter-District Road Infrastructure and Spatial Inequality in Rural Indonesia. *Economies*. 10. p. 229.

World Bank Transport and ICT. 2016. *Measuring Rural Access: Using New Technologies*. Washington DC: The World Bank.

Workman, R. and K. McPherson. 2020. *Measuring rural access for SDG 9.1.1*. Wokingham, UK: TRL Ltd.

Workman, R. et al. 2020. *Consolidation, Revision and Pilot Application of the Rural Access Index (RAI): Policy Guide and Roadmap to Achieve SDG Tier 1 Status*. London: ReCAP for Department for International Development.

Yu, J. et al. 2006. Development of speed related ride quality thresholds using international roughness index. *Transport Research Record*.

Zhao, G. et al. 2017. *Spatial and Temporal Characteristics of Road Networks and Urban Expansion*. Land, 6(2), 30. https://doi.org/10.3390/land6020030.

www.ingramcontent.com/pod-product-compliance
Lightning Source LLC
Chambersburg PA
CBHW041759220926
51847CB00014B/261
9789292775520